Edward Trevert

Electric Railway Engineering

Edward Trevert

Electric Railway Engineering

ISBN/EAN: 9783744649872

Printed in Europe, USA, Canada, Australia, Japan

Cover: Foto ©berggeist007 / pixelio.de

More available books at **www.hansebooks.com**

ELECTRIC RAILWAY
ENGINEERING

≈ BY ≈

Edward Trevert.

AUTHOR OF

EVERYBODY'S HAND-BOOK OF ELECTRICITY,
ELECTRICITY AND ITS RECENT APPLICATIONS,
EXPERIMENTAL ELECTRICITY,
A PRACTICAL TREATISE UPON ELECTRO-PLATING,
ARMATURE AND FIELD-MAGNET WINDING, &c., &c.

ILLUSTRATED.

EMBRACING PRACTICAL HINTS UPON POWER HOUSE, DYNAMO,
MOTOR AND LINE CONSTRUCTION FOR
THE USE OF STUDENTS.

LYNN, MASS.
BUBIER PUBLISHING COMPANY.
1892.

PREFACE.

ELECTRICITY is rapidly supplanting horse-power in our street car service, and it may not be many years before steam will give place to electricity on our railways. Already these indications are sufficient to be of interest, not only to the student in this department of electrical science, but, likewise,

ERRATA.

Alkaline zuicate cell, on page 106, should read alkaline zincate cell.

Potassium zuicate, on same page, should read potassium zincate.

Emery dust, on page 133, should read sand dust.

Wightman Single Railway, on page 184, should read Wightman Single Reduction Gear Railway.

Westinghouse four-pole single, on page 184, should read Westinghouse four-pole single reduction gear.

Edison slow speed single rod, on page 184, should read Edison slow speed single reduction gear.

Locomotives by heavy traction, on page 186, should read locomotives for heavy traction.

Causes that make it revolve, on page 182, should read causes which make them revolve.

leading electrical companies for some of the excellent illustrations contained herein.

EDWARD TREVERT.

LYNN, MASS., May 1, 1892.

PREFACE.

ELECTRICITY is rapidly supplanting horse-power in our street car service, and it may not be many years before steam will give place to electricity on our railways. Already these indications are sufficient to be of interest, not only to the student in this department of electrical science, but, likewise, to the general public. Much interest centres in the mechanism of such inventions, and in the modes of operating them.

In this book it has been my endeavor to make the subject as plain and interesting as the present advance in the science will admit. Illustrations have been inserted wherever the text could be made clearer by their use.

I have briefly referred to steam apparatus, as there are already many excellent works upon that subject, besides it would be impossible to do the subject justice in a treatise of this size. I have therefore written from a purely electrical standpoint, trusting that the reader will find enough to instruct and interest him in this branch of electrical science. If my efforts in this direction meet with approval, this book will have fulfilled its mission.

Some of the articles have been compiled from the leading electrical journals, such as *The Electrical World*, *The Electrical Engineer*, *The Electric Age*, *The Electrical Review* and *The Electrical Industries* to which I wish to express my sincere thanks. I am also indebted to the several leading electrical companies for some of the excellent illustrations contained herein.

EDWARD TREVERT.

LYNN, MASS., May 1, 1892.

CONTENTS.

	PAGE
INTRODUCTION	7

CHAPTER I.
THE POWER HOUSE AND ITS APPARATUS	11

CHAPTER II.
RAILWAY GENERATORS	15

CHAPTER III.
LINE CONSTRUCTION	26

CHAPTER IV.
ELECTRIC RAILWAY MOTORS	41

CHAPTER V.
RHEOSTATS	70

CHAPTER VI.
ELECTRIC HEATERS	75

CHAPTER VII.
TROLLEYS	79

CHAPTER VIII.
LOCOMOTIVES FOR HEAVY TRACTION	84

CHAPTER IX.
TRUCKS	92

CHAPTER X.
CAR WIRING	98

CHAPTER XI.
THE STORAGE BATTERY SYSTEM . . . 104

CHAPTER XII.
SOME ILLUSTRATIVE ROADS 110

CHAPTER XIII.
SOME GENERAL REMARKS FOR MOTOR MEN . 124

CHAPTER XIV.
SOME GENERAL REMARKS FOR STATION MEN 132

CHAPTER XV.
CONCLUSION . . . 135

APPENDICES.

PAGE

APPENDIX A.
CHRONOLOGICAL HISTORY OF THE ELECTRIC RAILWAY 145

APPENDIX B.
FENDERS 149

APPENDIX C.
METHODS OF ELECTRICALLY CONTROLLING STREET-CAR MOTORS 151

APPENDIX D.
RAPID TRANSIT 166

APPENDIX E.
ELECTRIC STREET RAILWAYS AS INVESTMENTS . 176

Electric Railway Engineering.

INTRODUCTION.

TWENTY YEARS ago the Electric Railway existed only in the imagination of the electrician. Today it is a reality. What twenty years more will bring forth in this direction remains to be seen. Two systems are now in operation, the trolley and the storage battery. The trolley, being the most successful, is the one in general use. In the trolley system the electrical circuit consists of two parts, the overhead and the ground circuit.

In distributing the current, the rails are grounded and form one side of the circuit. If they have a good electrical connection from one to the other through fish-plates already in position, they form a path of very low resistance. Where such connections are poor, the rails are reinforced by a continuous conductor running the entire length of the line. The other part of the circuit consists partly of a hard-drawn silicon bronze or copper contact-wire of small size, but great tensile strength, which is suspended 17 to 18 feet above the track. (A diagram of the circuit is shown in Fig 1.)

This contact-wire is termed the working-conductor, and is carried over the centre of the track, at the height named, on insulators supported by span-wires running across from pole to pole and provided with additional insulators at their ends, or else by brackets which extend from poles placed on the side or centre of the street. The size of this wire is independent of the number of cars operated, or the distance over which the line extends.

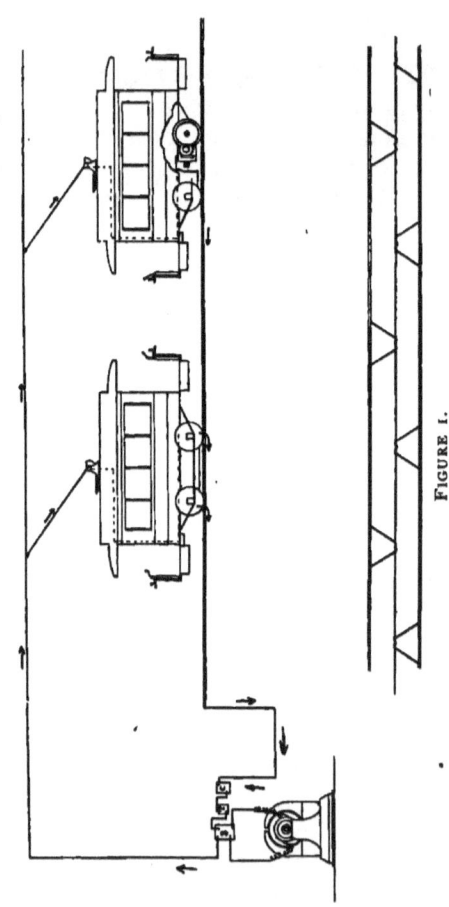

Figure 1.

The curves are formed of a series of short chords, which approximate the central line of curvature.

The whole structure is very light-looking, and it seems wonderful that it can be made the medium for the transmission of sufficient electrical power to propel any number of cars over any length of track.

The main current, however, is not carried by this small wire, but is carried by a main wire running parallel to the working-conductor and connected to it at intervals. This main conductor, which can be carried upon poles at the side of the street in the same manner as telegraph wires, or else buried in a subway, underground, is itself supplied at different points by feeders, which come from the main supply at the central station. Without the use of these feeders, which also can be reinforced if necessary, the size and cost of the overhead conductors would be very largely increased.

By means of these feeders, and by use of automatic cut-outs for dividing the line into sections, which are placed at the junctions of the feeders and main conductor, all danger of an accident on any portion of the line, disabling the operation of the remainder of the road, is avoided. This is a matter of great importance, and its value cannot be too highly estimated.

By means of this system, also, as the greater portion of the current is carried upon the main conductor and only a small portion on the working-conductor over the middle of the street, there is no change in the size of the working-conductor, and consequently no stoppage of travel required, with an increase in number of cars run or an extension of the line.

The current is taken from this working-conductor by a small structure on top of the car. This consists of a light trolley-pole supported upon a stout spring, so that it can move in every direction, and having at its upper end a grooved wheel, making a running flexible contact on the under side of the working-conductor. The flexibility of this arrangement is very great, it being able to follow with facility variations of the trolley-wire four or five feet in either a horizontal direction, or more than twelve feet in a vertical direction.

By this means a constant contact is made by the trolley-wheel at different rates of speed or around curves, and for different heights of the trolley-wire.

It is impossible (working underneath) to pull the trolley-wire down; and if off the line, the trolley can be replaced quickly and easily, even in the darkest night.

By the use of this underneath contact, not only are there no complicated switches on the overhead conductors, but all changing of contact is avoided when passing the turnouts.

In the storage battery system accumulator cells, or storage batteries, are charged at a power house, then placed upon an electrically equipped motor car (generally underneath the seats), and when fully charged will run twelve hours, after which time they are replaced by newly charged cells.

In the construction of the electric motor for car propulsion, the motor acts simply for the transformation of electrical energy with mechanical energy. A current of electricity is sent through the armature and field-magnets of the motor, which causes the armature to revolve. At present there are two classes of electric motors, fast speed and slow speed. With the fast speed motor the armature revolves with great rapidity, and the motion is communicated to the axle of the car by means of gears and pinions. In the slow speed motors the intermediate gears and pinions are left out, there being only one gear and pinion, the gear upon the axle of the car, and the pinion upon the shaft of the motor. To this class also belongs the gearless motor, whereby the motion is communicated directly from the armature shaft to the axle of the car, explanation of which will be made later on.

The electric railway may be divided into three parts: 1st, the power station and its apparatus; 2d, the line, and 3d, the electric motor and other car equipments. These we will now take up in their regular order.

CHAPTER I.

THE POWER STATION AND ITS APPARATUS.

THE equipment of a power station comprises the steam plant and the electrical apparatus. The steam plant consists of the boilers, engines, etc. The steam engines furnish the mechanical energy to drive the dynamos, and there are numerous makes especially designed for this purpose, for information of which the author refers the reader to the standard works upon steam engines.

The electrical apparatus consists of dynamos, voltmeters, ammeters, feeder boards, switch boards, lightning arresters, circuit breakers, etc. The general arrangement of a power station is shown in Fig. 2.

Supposing that the reader is already familiar with the general construction of the dynamo, it need not be described here.

The arrangement of a switch board is shown in Fig. 3.

The feeder, or connection board, consists of metal connections, in the circuit of which are interposed fuses. These fuses are strips of metal so proportioned as to carry the right amount of current. An excess of current causes them to melt or blow out, thus breaking the circuit and preventing damage to the electrical apparatus. The automatic circuit breaker is an apparatus operated by an electro-magnet and a powerful spring which throws open a switch. When the current exceeds a normal amount the magnet acts upon the armature, releasing the switch, which is thrown open by springs. The voltmeter consists of a needle acted upon by an electro-magnet. This needle is made to traverse a scale which is graduated into degrees, representing so many volts. From this instrument the electro-motive force is determined.

The lightning arrester is an apparatus used to carry away lightning

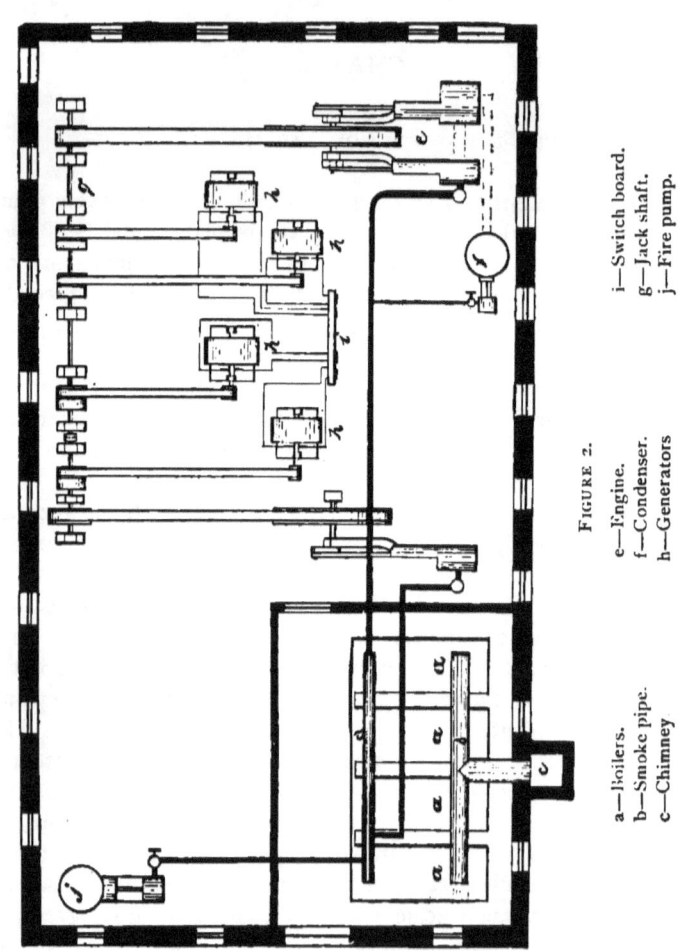

FIGURE 2.
a—Boilers.
b—Smoke pipe.
c—Chimney
e—Engine.
f—Condenser.
h—Generators
i—Switch board.
g—Jack shaft.
j—Fire pump.

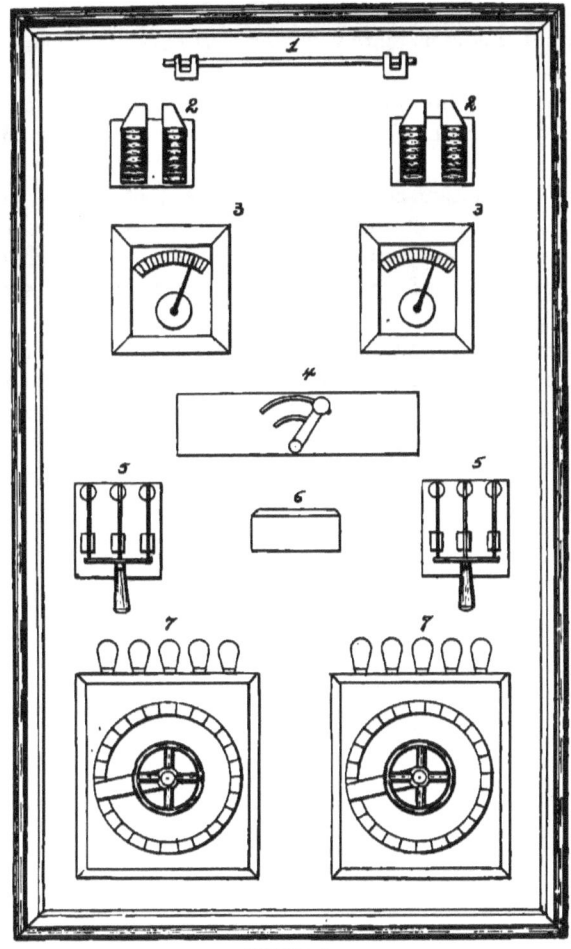

FIGURE 2.

1—Equalizer.
2—Circuit Breaker.
3—Ammeter.
4—Potential Board.
5—Switch.
6—Voltmeter.
7—Rheostat.

discharges and prevent them from doing damage. If lightning strikes the circuit it is immediately carried to the ground through this instrument.

The ammeter is an instrument similar to the voltmeter, only it gives the amount of current in the circuit, said amount being measured in amperes.

Dynamos or generators come under the head of power house apparatus, but on account of the large amount of space necessary for their description they will be taken up in the next chapter. There is a large variety of them, all of which have their special claims to superiority.

As the dynamo is the head of the electric railway, too much care cannot be taken in selecting one for its operation. All the machines described in this book are of first-class manufacture and made by standard companies.

CHAPTER II.

RAILWAY GENERATORS.

SPECIAL dynamos are now constructed for the generation of the electric current to operate railway motors. Experience has taught electricians that it is much more economical to use large generators. They are usually wound to give a large current at a pressure of 500 to 600 volts. For example: One of the large generators used to furnish power for the West End Railway of Boston has an output of 250,000 watts, which is equal to nearly 300 horse-power. It is a multipolar, that is, it has four pole pieces. Its armature is a Gramme ring, its commutator has 180 sections, and the armature is run at a speed of only 400 revolutions per minute. Ring armatures are used almost exclusively in railway generators, one of the principal reasons being that if a coil burns out or is in any way damaged it can easily be removed for repair.

The field magnets of dynamos for railway work are usually compound wound in order to secure perfect regulation and to obtain a constant current.

It is now proven that in winding a magnet it makes no difference so far as the magnetic effect is concerned, whether there are 20 turns of wire, with 5 amperes flowing through them, or 5 turns with 20 amperes, the result is the same. The product of the number of turns by amperes is called the "ampere turns." In a series wound dynamo the whole current is carried through the field coils which are connected in series with the armature and external circuit. This machine is used almost exclusively for arc lighting. (A diagram of a series wound dynamo is shown in Fig. 4.)

In the shunt wound dynamos the field magnets get their current at nearly constant potential, they being wound with fine wire to give them a high resistance, and the small amount of current being replaced by a large number of turns (See Fig. 5).

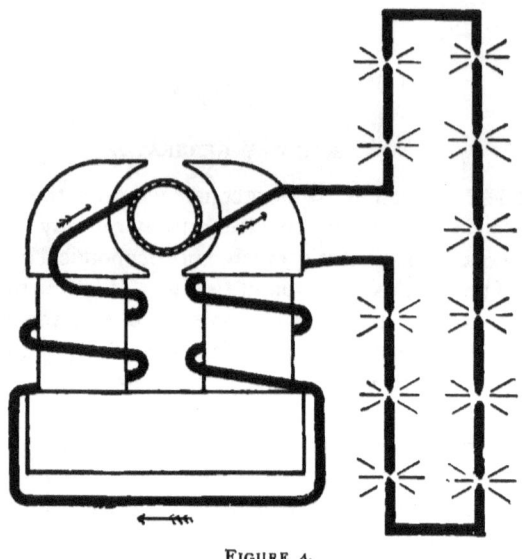

FIGURE 4.

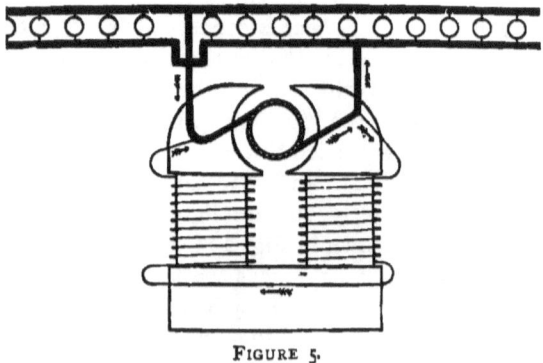

FIGURE 5.

In the compound wound dynamos the field magnets are wound in both series and shunt, by which methods an absolute constant potential is obtained. (See Fig. 6.)

For more detailed information on the subject of winding the reader is referred to the author's book, "Armature and Field-Magnet Winding." Many smaller differences of mechanical and electrical construction are adopted by the several electrical companies, which can best be understood by a brief description of their machines, to which now let us devote our attention.

An illustration of the *Short Electric Railway Generator*, of 150 horse power, is shown in Fig 7. It is a very massive machine and is

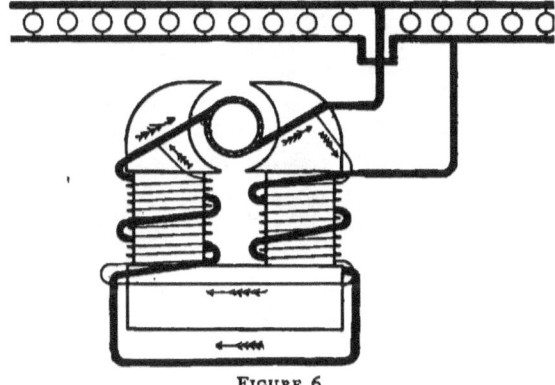

FIGURE 6.

capable of delivering a current of 225 amperes at a pressure of 500 volts. The field magnet frame is one large casting, weighing over 3,000 pounds, of soft iron slowly annealed. To this are bolted eight field magnets, carrying the shunt and series coils, and provided with pole pieces of peculiar shape, arranged for side presentation to the armature. The armature of this generator possesses distinctive characteristics. It is of the Gramme ring construction. The massive spider carrying the foundation ring upon which the armature is built, is keyed to a shaft nine feet long and six inches in diameter. The

FIGURE 7.

armature core is formed of thin sheet iron wound spirally on the foundation. By this method of winding each of the 200 coils is exposed to the air on all sides, thus receiving perfect ventilation. The diameter of the armature is 36 inches. Another feature is in the commutator box, where there is an adjustable ball bearing thrust collar containing several hundred balls, and so arranged as to carry the armature thrust in either direction without heating. The commutator has 200 segments, so that the pressure between adjacent segments is unusually small and there is no sparking. There are four brushes, which are held together by two independent collars and sets of brush holders.

The Thomson-Houston 300 H. P. Railway Generator is a multipolar dynamo and the fields are compound wound (see Fig. 8) and has an output of 250,000 watts, which is equal to about 300 h. p. The armature is of the Gramme ring pattern and so constructed that opportunity is afforded for the best insulation, and the danger due to great difference of potential between any two of its conductors is avoided. This is a most valuable and important feature, as in case of accident or injury to any coil, it can be easily repaired without affecting in any way the remaining coils. The construction of the armature affords excellent ventilation, which is very necessary in dynamo machines, particularly as their size is increased, for the reason that the radiating surface does not increase in proportion to the size of the mass.

In order that the conductors inside the armature may be held securely in place, an adjustable internal wire support has been designed. When the armature is being wound the wires are forced into position so that they cannot sag, vibrate, or chafe the insulation. All tendency to short circuiting is thereby avoided and the position of the wires assured.

The commutator has 180 sections. In practice, the generator will have its fields separately excited, although the connection at the switch-board is so arranged that by throwing a switch the dynamo can be made self-exciting, should emergency require it.

FIGURE 8.

The movement of the brushes is affected by means of the shaft, on which a small worm is attached, and which in turn works in a rack fastened to the yoke. By means of this a very fine adjustment of the brushes can be made. The worm locks the yoke so that it cannot be moved except by hand.

One of the most important features of this generator is the arrangement for lubrication and good alignment of the bearings. The boxes are made in two parts and are entirely separate from the stands. On the top of the stand is a seat.

The total floor space occupied by the 300 h. p. generator is 13 ft. 3 1-2 in. x 7 ft. 1 in. The height of the machine is a little less than 8 ft. The pulley is 43 in. in diameter and has a 35 in. face. The speed is 400 revolutions per minute and the dynamo, complete, weighs about 21 tons.

The Mather Electric Railway Generator. — Recognizing the demand for power transmission by means of the electric current, the Mather Electric Company has brought out a series of machines for that purpose. The generators are built up to 30,000, 50,000 and 75,000 watts with four poles, and 180,000 watts with six poles. Drum armatures are used in all the machines. In the four-pole machines the winding is such that the current has but two paths through the armature wires, and by a special method, devised by Prof. Anthony, no two wires having any great difference of potential are brought near each other.

Fig. 9 represents the 75,000-watt generator, showing the general character of all the four-pole machines, with the field magnet in one casting. In the 180,000-watt six-pole machine the field magnet is cast in two halves, but divided through the middle of two opposite poles instead of across the magnetic circle.

The New Multipolar Generator, made by the Westinghouse Electric & Manufacturing Company, of Pittsburg, Pa., is shown in Fig 10. In this machine the pole pieces project radially from the

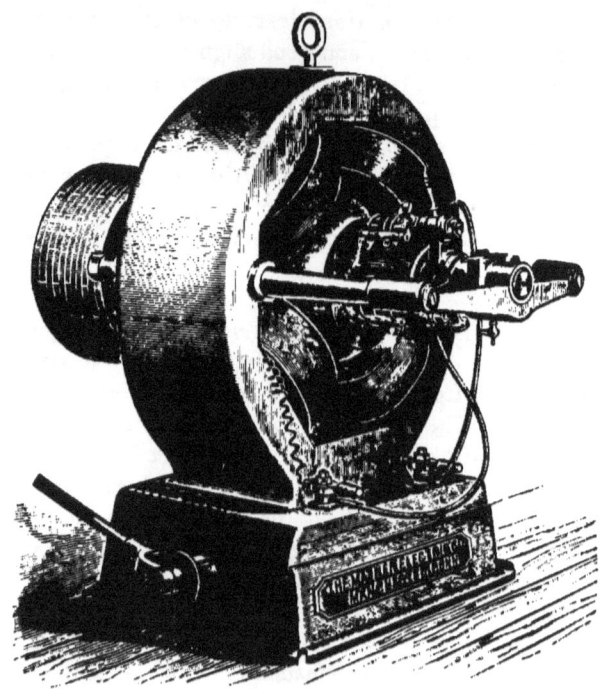

FIGURE 9.

interior of the cylindrical yoke pieces, and by the peculiar construction ready access may be had to the field coils and armature. The machines are all wound for 500 volts, E. M. F., but by means of a rheostat this can be raised to 550 or 600 volts. They are self-exciting and compound wound machines. The armature is a distinctive feature. It is of the Siemens' type, the core of which is built up in the usual way, of a large number of thin iron discs which are rigidly keyed to the shaft. The wires are not placed on the exterior of the core, as is usually done, but are placed in insulating tubes which are imbedded in the iron of the core. This construction obviates the

FIGURE 10.

use of binding wires. A special method of winding is used, and the amount of wire necessary is reduced to a minimum. The commutators are long and massive. The brush holders are composed of independent holders, thus allowing each carbon brush to be removed without disturbing the others. The machine is carefully regulated, designed for railway use and to require a minimum amount of attention.

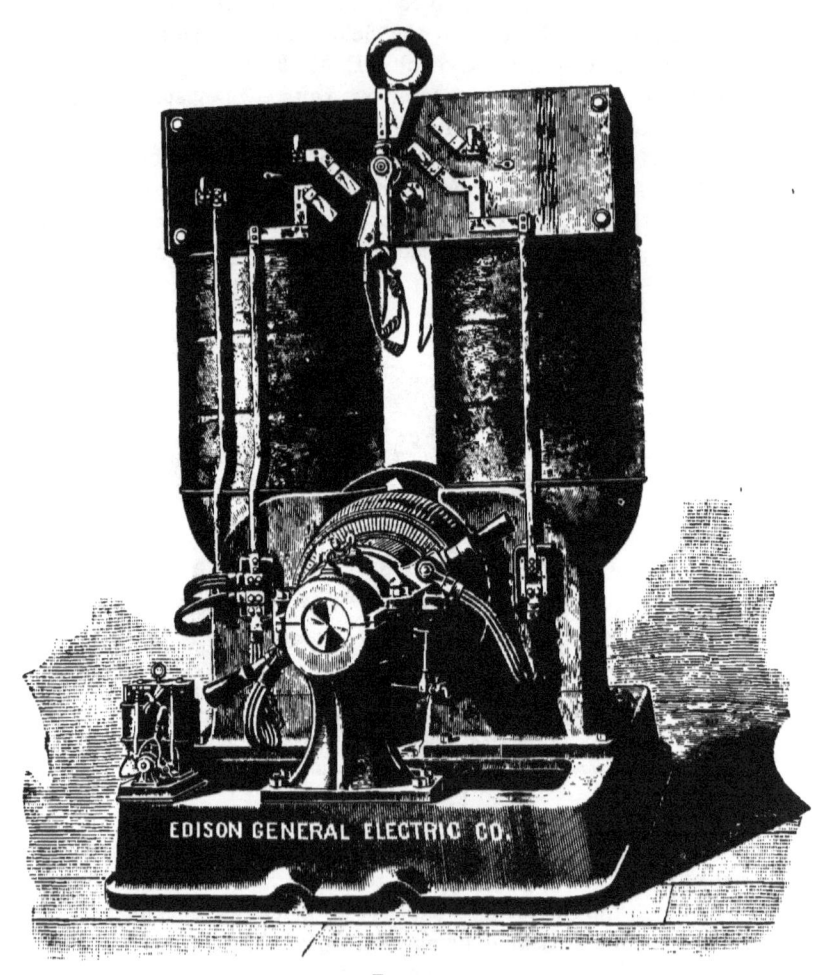

FIGURE 11.

Two Hundred Kilo-Watt Edison Generator. — In the accompanying illustration (see Fig. 11), is shown the latest form of Edison railway generator of 200 kilo-watts capacity. As will be seen, it is of standard bi-polar type, the general features of which are so well known as to need no further description. To adapt this generator to the demands of electric railway service, its field has been supplied with a compound winding, easily adjustable to meet the necessary requirements by means of a shunt coil, which is conveniently placed in the back board of the keeper. The close adjustment obtained by this arrangement greatly facilitates the operation of generators in parallel, and forms one of the characteristic features of this particular type. The series field is composed of sections wound on spools, which are slipped separately over the cores, and then properly connected. In the event of a fault occuring, the spool in which it develops can be removed and another substituted at once; this not only prevents delay, but makes any repairs necessary a matter of comparatively small expense. The armature is so wound that it has two distinct windings, and each end is furnished with its commutator, rocker-arm and brush holders. The centre of gravity of the armature is low, due to the bearings being located to the base frame, great stability is secured. Self-oiling bearings and carbon brushes help to reduce to a minimum the attention necessary to the operation of a dynamo.

CHAPTER III.

LINE CONSTRUCTION.

THE "overhead trolley" system of electric railroads is undoubtedly the best in active, practical use in this country. The wire from which the current is taken is suspended over the centre of the track, about fifteen feet above the rails. A long arm or pole, called the trolley, hinged upon the car roof is raised by springs, so that a shoe or wheel at the outer end is pressed upward against the lower surface of the "trolley wire."

The current from the dynamos at the generating station comes through this wire, down the trolley arm, through the motors to the rails, thence back to the dynamos to make the circuit complete.

In discussing the system it will be well to consider it under several sub-divisions.

1. Supports { Poles. / Brackets.
2. Cross Wires for Suspension.
3. Insulators or Hangers { Single Insulation. / Double Insulation.
4. Ears { Clamping. / Soldered.
5. Switches, Frogs, and Crosses.
6. Sections and Feeders.
7. Guard Wires.
8. Roadbed and Rails.
9. Return Circuit (Ground).
10. Double System.

1st. When the ordinary pole construction is to be used, the poles should be set at the very outset, that the soil may get securely packed about them before the real strain on the wires begins. Poles at least eight inches diameter or square should be used. The first roads equipped used weaker supports, which bent and sprung so as to loosen the cross wires and allow the trolley wire to sway and vibrate. When first erected the poles on opposite sides of the street should incline from each other, so that by the time the cross wires are of the necessary tightness, the poles will stand erect. (Fig 12.) If this precaution is observed and the poles are set six to eight feet deep, well packed with flat stones, the construction will require no repairs for ten years. It is well to coat the lower ends

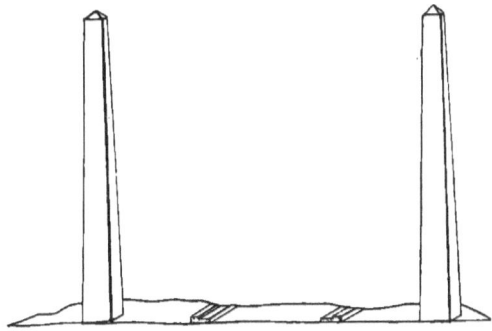

FIGURE 12.

of the poles with tar to prevent rotting. Iron poles made by different sizes of pipes slid into each other, or girder iron is used where exceptional strains are to be borne.

The arrangement of poles on opposite sides of the street is desirable when the tracks are in the centre of the street. If the rails are close to one side the "Bracket Suspension" is the neatest and most desirable. Only one line of posts is needed, and on these are fastened brackets with arms about six feet long, reaching to the centre of the track. The arms of the brackets are made of one and one-half inch gas pipe and thus combine lightness with rigidity. (See Fig. 13.)

28   ELECTRIC RAILWAY ENGINEERING.

In the case of a double track road, it is sometimes convenient to use a line of posts erected between the tracks, with double brackets, reaching to the centre of both tracks, one piece of pipe extending through the post to overhang the tracks on both sides. (See Fig. 14.)

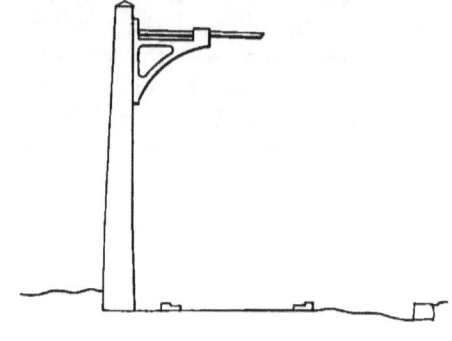

FIGURE 13.

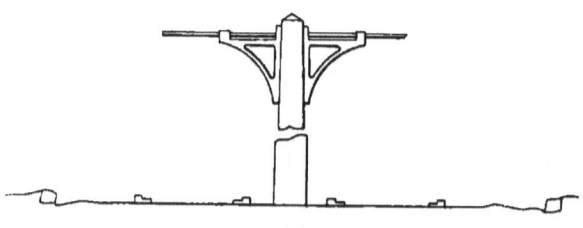

FIGURE 14.

2d. The cross wires are usually of galvanized iron about three-sixteenths of an inch in diameter. At one post the end can be secured in a screw eye or an eye bolt; the other end should be fastened to an eye bolt with shank long enough to pull the wire tight, and also to take up slack wire after a period of use. (Fig. 15 shows the arrangement.)

Depending on the kind of insulators used, the cross wires are sometimes one continuous piece, at other times they are made in two pieces (three for double tracks), separated where fastened to the insulator. Sometimes, still another parting is made to insert a lightning arrester.

For supporting frogs or other heavy pieces incident to the system, stronger and even double wires are necessary. "Anchoring and

FIGURE 15.

FIGURE 16.

FIGURE 17.

strain wires" are of the nature of cross wires. Where section insulators (described later) are inserted, diagonal strain wires are necessary to relieve the somewhat weak structure of the insulating material.

3d. The trolley wire must be well insulated from all electrical connection with the earth. The poles, though usually of wood, cannot be depended upon to answer this requirement. The simplest and earliest method was to insert an ordinary white glass insulator

30 ELECTRIC RAILWAY ENGINEERING.

in the cross wire on each side of the centre, or place where the trolley wire was located. (See Fig. 16.)

There is a short section of wire between the two insulators, from which the trolley wire can be suspended. Such insulators are

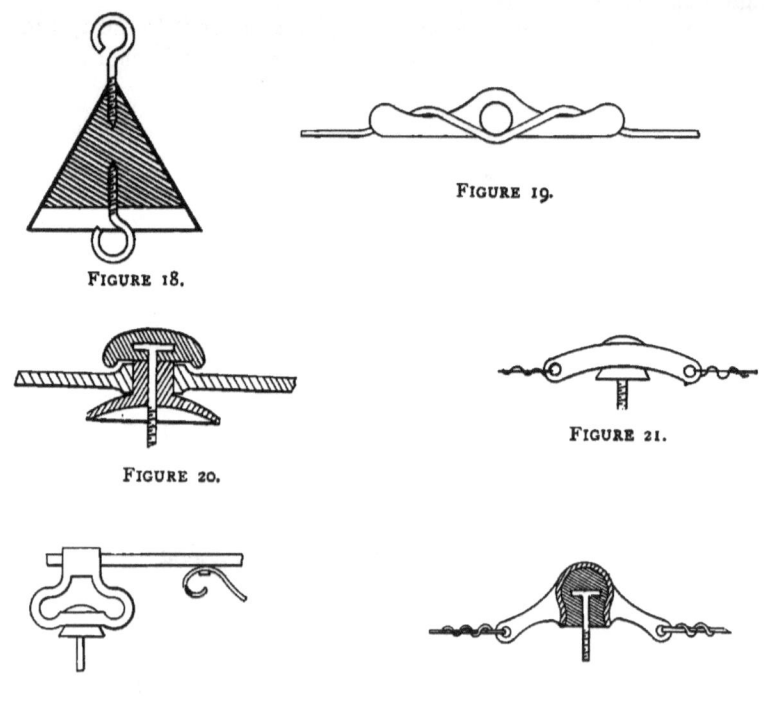

FIGURE 18.

FIGURE 19.

FIGURE 20.

FIGURE 21.

FIGURE 22.

FIGURE 23.

not much used, however, on account of their frequent breakage, and poor insulating properties, when covered with moisture. Sometimes a double insulator, consisting of two straps of iron, rivetted together with the glass between, is practicable. (See Fig. 17.)

Such a device is usually employed in some portion of a strain wire. A wooden cone in a sheet metal case makes a tolerably good insulator. A screw eye in the apex, another in the base, affords means for suspending itself and the trolley wire to the cross wire. (See Fig. 18.)

Mechanically, this construction is somewhat weak. An excellent form is shown in Fig. 19. It is made of cast-iron with a hole in the centre about one and one-fourth inches in diameter. Projecting arms on each side are cast with a twist and groove, so that the cross

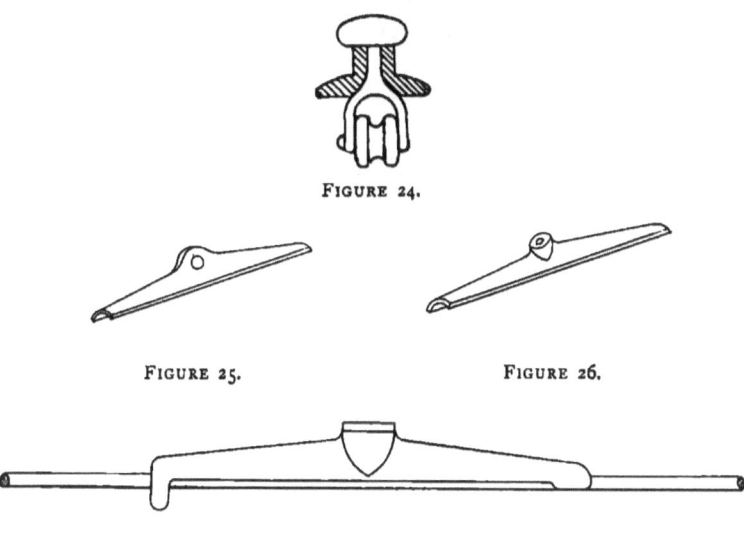

FIGURE 24.

FIGURE 25.   FIGURE 26.

FIGURE 27.

wire can be bent under and over in an S-shaped curve. Once in position it is not easily dislodged.

A bushing and cap of moulded insulating material, such as mica and shellac, or rubber and clay, fills the hole and allows for a bolt (described later). Fig. 20 shows one construction.

This form of insulation is also used in the "arch and Bracket suspension." (Figs. 21 and 22.)

A form of insulator and holder considerably used consists of an inverted cup of iron, provided with ears for attaching the cross wires. A large headed bolt is held in the cup by rosin or sulphur cast to fill the intervening space. (Fig. 23.)

Double insulation as applied to line construction consists in insulating the metal body of the holder from the cross wires. Ordinary porcelain knobs set into forked ends in the arms of the holders are used. (Fig. 24.)

On curves the trolley wire needs to be drawn tight from one side only and the holders need lugs for attaching the cross wires only on one side.

4th. The ears are the parts to which the copper trolley wire is

FIGURE 28.

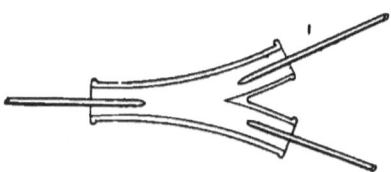

FIGURE 29.

directly attached. They are connected with the insulator above and hold up the wire beneath. Two forms are commonly used: those rigidly attached, usually soldered, or those only clamped and hence adjustable. The former are made of cast brass with a lug projecting upward to catch on a hook of the insulator or to screw upon a bolt, (shown in Figs. 18 and 20). The lower edge of the ear is grooved to receive the trolley wire. The size of wire usually employed is five-sixteenths of an inch in diameter. The ear is about a foot long and is carefully soldered its entire length, to the wire. (See Figs. 25 and 26).

On account of the sagging of the wire between the places of support, the solder is apt to loosen at the ends of the ears; for this reason the ends are usually very slim, so as to bend easily with the wire; sometimes projections are cast on the ends that are bent around the wire and soldered as shown in Fig. 27, one end showing the shape before soldering.

Clamping ears are made in two pieces, tightened against each other with the trolley wire between. These have this advantage that when the poles bend or change their position or the wire sags or stretches they can be moved to correspond. They have this disadvantage that being more bunchy than the soldered ears the trolley wheel is more likely to rebound when striking them, and draw a flash of fire that burns both wheel and wire. (Fig. 28.)

The weight and motion of the wire sometimes bends down the thin edges of the ears, and allows the line to fall into the street. A modification of this form has been used, consisting of a brass body somewhat like the soldered ear, but having a piece of sheet copper wrapped around under the trolley wire and fastened with screws or rivets. The trolley wheel flashes at every passage across the sheet metal and soon destroys itself.

5th. Switches, frogs and crosses for the trolley wire are located directly over the correspondingly named portions of the track. They are made of cast brass trough-shaped with the wire entering the centre of the opening. The trolley wheel runs along under the wire directly into the frog, where the wire no longer has any directive power, the walls of the trough keeping the wheel in place. On leaving the frog the trolley wheel will be drawn into the trough corresponding to that track which the car has just entered. "Two-way," "three-way," switches or frogs and acute and right angle crosses are shown in Figs. 29, 30, 31 and 32.

The trolley wire itself passes on top of the brass body of the frogs or switches, and supports them by means of bolts and clamps.

6th. At first thought, it would seem desirable that the overhead wire be strung in one continuous length. If this were done, however, an accident such as the falling of any part of the line, would necessitate a shutting down of the whole system. It is well to divide the road into portions or sections and supply each independently with electricity. Then in case of accident it is possible still to keep the greater part of the road in operation. Some electricians advocate sections five hundred feet long, others a quarter or half a mile

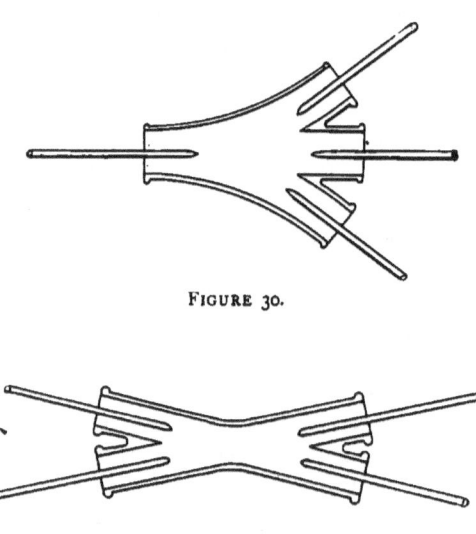

FIGURE 30.

FIGURE 31.

in length. Of course the break between two sections must be small in order that the trolley wheel may have a continuous path. Two ears, into each of which one end of the wire is attached, are held close to each other by an insulated iron ring bent in a diamond shape. The horns or ends of the ears are allowed to approach each other to a distance of about three-eighths of an inch. (See Fig. 33.)

LINE CONSTRUCTION. 35

A flash usually follows the passage of the wheel across the break, and lately it has become customary to make several small breaks or separations rather than a single large one.

A separate supply wire or "feeder" from the switch board at the generating station is run to each section; fuses or switches control the amount of current proper for the demands of each. Thus it is possible to shut down or start up any part of the road by shifting switches at the power house. The feeders are attached to poles

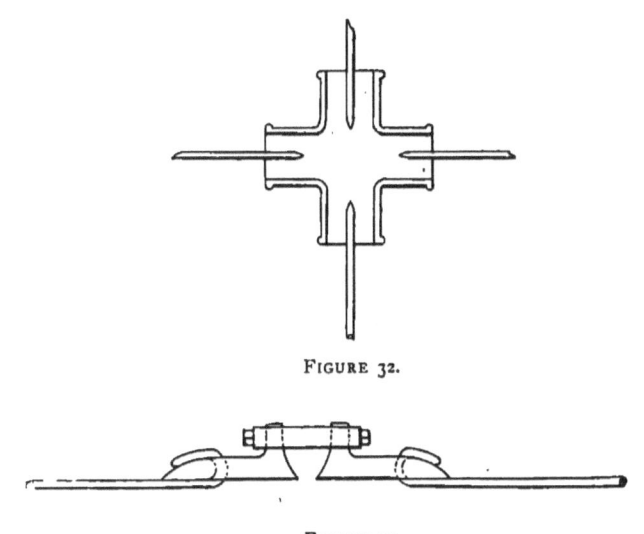

FIGURE 32.

FIGURE 33.

along the route and at proper intervals branches are let out along the cross wires and attached to the ears that support the trolley wire.

7th. Should telegraph, telephone, or other wires fall across the trolley wire a part of the current would be diverted, and besides causing some danger to operatives, would melt the delicate instruments connected with the wires. As a precaution against this emergency, franchises for the construction of the electric roads usually call for iron "guard wires" to be stretched above and parallel with the trolley wire. Any falling wires will then be intercepted. Sometimes these guard wires are strung on insulators that are on horn-like arms extending from the trolley wire hangers.

It is better to get these wires some distance from the trolley wire, say, two or three feet. To accomplish this, the regular cross wires should be attached to the poles, not at the upmost point, but a few feet from the top; above these there will then be room for other cross wires whose whole duty will be to support the guards.

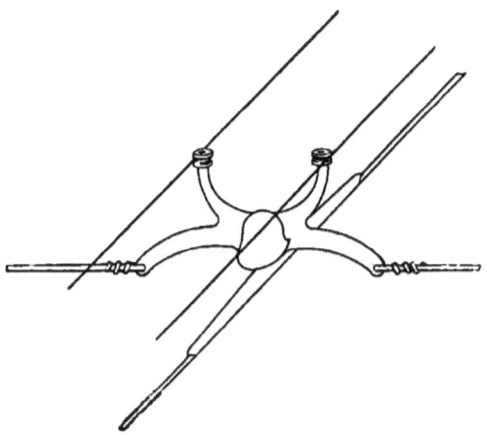

FIGURE 34.

8th. Electric cars are very heavy, for their size; but heavy weights, if supported on springs, can easily be carried on tracks without serious shocks to the rails. Electric motors under cars are partly suspended on an elastic cushion or spring. The greater part of the weight, however, is hung directly on the axle. The whole inertia of the heavy motors is carried to the rails. As the wheels pass from one rail to the next the shock amounts to a sledge hammer blow, and the ordinary rails used on horse car lines are quite unsuitable for electric railway use.

9th. The rails for electric lines should be much like those for steam cars. They must be stiff enough to resist appreciable bending as the car moves along, and have enough body of metal to prevent their edges and ends from losing their shape by the hammering strains. Electric cars have been run over existing horse traction lines, but the result has been that the tracks have become loosened from the ties and crushed out of shape at the joints. A construction

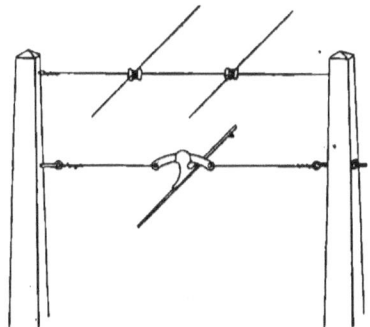

FIGURE 35.

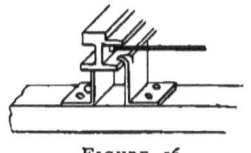

FIGURE 36.

similar to such roads but with considerably heavier materials makes a very fair road, but what is called the "girder rail" with chair supports attached to cross ties, is without doubt the very best. The rails are deep and rest on brackets of wrought iron, which in turn are nailed to ordinary cross ties. The result is a track that depends for its support on parts covered deeply with gravel or paving stones and which cannot yield or loosen. (See Fig. 36.)

The rods held by nuts at each end are of course to be used without stint.

Figure 37 shows a plan of *Single Curve Overhead Construction.*
Figure 38 shows a plan of *Double Curve Overhead Construction.*

10th. Thus far attention has been paid to the means of getting the power to the moving cars. The electricity must have a return in

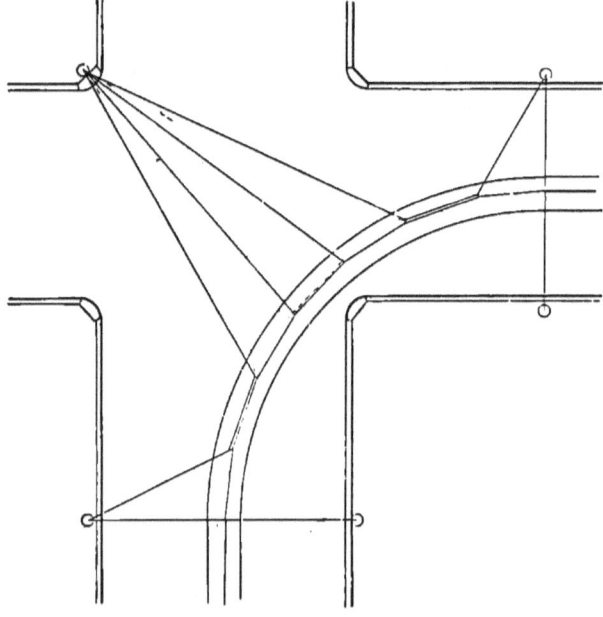

FIGURE 37.

order to complete its circuit; and after doing its work is expected to get back to the dynamos without further trouble or expense; but the cheapness of obtaining this return depends upon certain conditions. All the rails must be carefully connected together "electrically." Mere mechanical contact at the "fish plates" is nothing to be depended upon, as these parts are very rusty and often loose. Copper wire must be rivetted or soldered to the rail to connect it with its neighbor. It is well to lay a copper wire in the earth parallel with the tracks, connected to every rail. Then in case one or more rails are removed for renewal or repairs, the electric circuit will not be interrupted. At the generating station heavy wires are connected

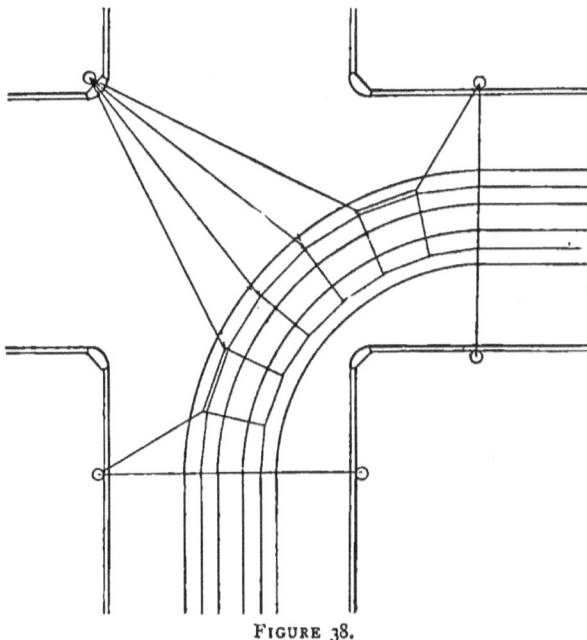

FIGURE 38.

with the rails and with the earth wire, these leading to the negative side of the switch-board.

11th. Sometimes objections are raised against using the ground or rails for a return circuit. These objections are usually sentimental or imaginary. A substantial one is urged by owners of telephone lines, as the induction from a single trolley wire causes disturbances in the telephone receivers. This induction is due to this fact, that the line-wire of both telephone and railroad systems are near and parallel to each other and both use a ground return. The evil can be remedied by either systems being equipped without-going and return wires close to each other. It is easy for the telephone companies to do this, and indeed for long distance transmission is always necessary.

A double overhead trolley system is complicated and dangerous; the wires having the whole difference of potential are close to each other, and difficult to keep apart. Switches and frogs make an almost interminable network. In the single trolley system, the current passing from the wheels to the rails causes such a heating or welding action at the points of contact that the traction is considerably increased over the amount due to mere friction. In the double system this beneficial advantage is not attainable, and thus aside from the cost of the insulation compares unfavorably with the other and simpler system.

CHAPTER IV.

ELECTRIC RAILWAY MOTORS.

IN the construction of the electric motor for car propulsion, the motor acts simply for the transformation of electrical energy with mechanical energy. A current of electricity is sent through the armature and field magnets of a motor which causes the armature to revolve. Formerly fast speed motors were used in electric railway service, in which the armature revolved with great rapidity, necessitating the use of numerous gears and pinions by which the motion was communicated to the axle of the car. At present, slow speed motors are almost wholly used, by which the intermediate gears and pinions are left out, there being only one gear and pinion, the gear being upon the axle of the car and the pinion upon the shaft of the motor. There are, however, some exceptions to the rule. These exceptions are in gearless motors, particulars of which will be given later in this chapter. We will now call attention to the different styles of railway motors.

The Wightman Single Reduction Railway Motor.—Among the first to recognize the desirability of, as well as the possibility of, eliminating one set of transmission gears in electric street railway cars was Mr. Merle J. Wightman, who, as electrician of the Wightman Electric Manufacturing Co., of Scranton, Pa., over a year ago commenced experiments toward the development of a slow-speed single reduction motor. The results of this work are embodied in the motor shown in the accompanying engraving, Fig. 39, from which it will be seen that the "Kennedy" type of field-magnet is employed. This form of field-magnet has the advantage of almost completely covering the field coils and producing an "iron-clad" motor.

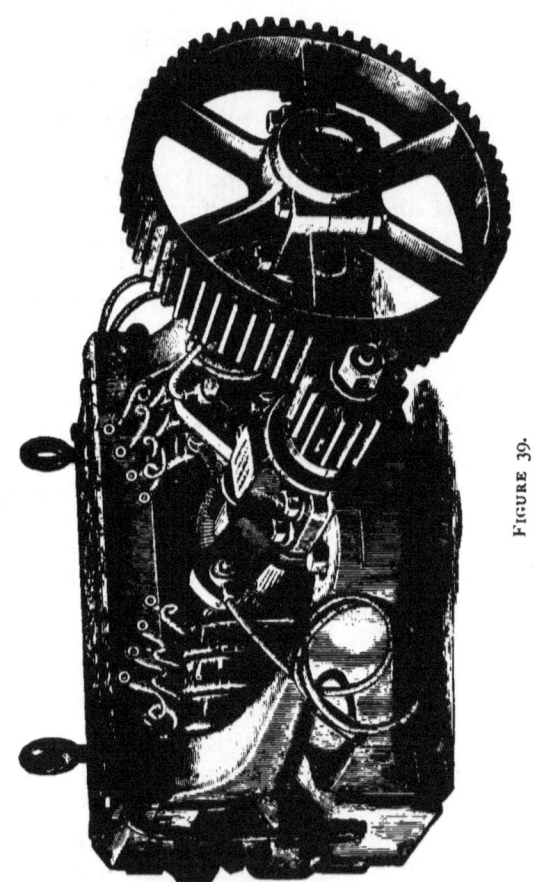

FIGURE 39.

It gives a very strong and efficient field and all four poles are excited by two field windings.

The armature is of the Gramme type (see Fig. 42), and the commutator is cross-connected so that but two brushes are used, placed at an angle of 90 degrees and on top of the commutator.

The cross-connecting of the commutator is accomplished in a remarkably simple way. All the crossing cables are formed symmetrically into a flat disc which is firmly bolted to the head of the commutator and becomes an integral part of it. In this way all possibility of vibration and risk of breakage is overcome. The commutator lead-wires are all flexible cable after the Wightman Company's

FIGURE 40.

well-known method of armature winding. These lead-wires are fastened to the commutator without screws and in such a way that they can be detached in a few minutes, when it becomes necessary to remove a commutator. The armature is mounted within a strong, continuous frame forming part of the field-magnets. The bearings are self-oiling and dust-proof and are designed to be used with grease, oil, or both.

Either field winding is removable without disturbing the other or the armature, each winding being made up of separate coils, one of which is shown in Fig. 40. The removal of two bolts at one end makes it possible to lift out one of the fields, after which the armature can be taken out. The top field pole is hinged at one end for convenience in removing the field, or armature. The ratio of the reduction of the gearing is 4.4 to 1, the armature pinion having

FIGURE 41.

fifteen teeth and a diameter of five inches. This ratio gives about 480 revolutions of the armature at a car-speed of 10 miles an hour.

The aim of the designer of Wightman motor has been to attain as great an efficiency as possible with the wide variation of speed and load met with in street railway practice. This has been obtained by means of large field magnets of a great number of turns of wire. In fact, speed regulation is obtained without the use of any external resistance above three or four miles an hour. On a level, cars equipped with two 20 h. p. Wightman motors have frequently attained a speed above twenty-five miles an hour.

Mr. Wightman's experience has lead him to the belief that there is no economy in operating motors of small capacity. Many roads are operated in such a way that cars are barely maintained on schedule time by dangerous and reckless running on down grades. A little calculation will show that by the expenditure of a little more power grades may be climbed rapidly, and as a result, much more service can be gotten from a given expenditure in wages for conductors and motor men and interest on plant; and the cost of the extra coal will be comparatively insignificant. It is much safer to climb grades rapidly rather than to descend them at a high rate of speed, not to mention the greater satisfaction of patrons. When climbing a grade a stoppage of power and application of brakes will bring a car to a standstill within surprisingly short distance. Since the wear and tear of ample-sized motors is obviously less than those overworked, all considerations of economy and safety would therefore point to the use of the former.

While in the Wightman motor electrical perfection has not been sought for at the expense of simplicity and durability, a very high efficiency is obtained. The armature resistance of the 20 h. p. motor is .75 ohms, and that of the main field coils .15 ohms, with a load of 40 amperes, or over 26 electrical horse-power; this would give a loss of potential in the motor of 36 volts, or an electrical efficiency of 92.8. Even with this excessive load the commercial efficiency has been found to be as high as 87 per cent. The large field, referred to above, makes possible a high efficiency at low speed and light loads. These qualities are synonymous with powerful torque or startling force. A loaded car equipped with Wightman motors requires not more than from 15 to 20 amperes to start on a level.

Not the least interesting improvement in car equipment is the new controlling device employed by the Wightman Co., shown in perspective in Fig. 41. Here, again, simplicity and durability have been the aim of the designer. Corresponding points in each controller are connected at each end of the car, and all mechanical contrivances beneath the car, such as reversing switches, rheostat cables, are done away with. There are five speed contacts on each side of the middle stop. A movement of the controller handle to the left causes the car to go forward, while an opposite movement reverses the direction of motion. The gradation of resistance on the reversing side of the controller is such that the car can be brought either slowly or suddenly to a standstill without the use of brakes or undue strain on the

FIGURE 42.

motors. The control is as absolute and flexible as in the case of a steam locomotive, yet very much more convenient in operation. The top of the controller is provided with notches in which a catch on the operating handle engages. This arrangement enables the motorman to confine his attention to the track ahead, and yet be aware of the position of the controlling lever.

Another valuable feature of the controller is the device for extinguishing the arc formed on breaking contact. This consists of a small magnet let in from the back of the slate base and the poles of which come directly opposite the spaces between the contacts; the well-known action of the magnet serves to blow out the arc and thus preserves the contacts. The latter in addition are all arranged to be readily removable for renewal, in case of necessity, and for this purpose all connections are in sight and can be gotten at merely by removing the cover.

The Thomson-Houston Single-Reduction Gear Railway Motor.— An examination of Fig. 44 will show that the motor is practically iron clad, having two internal pole pieces of ample spread, carrying

FIGURE 43.

two field spools which practically surround the armature core, on the same principle as the well-known arc dynamo, designed by Prof. Thomson. The magnetic circuit is completed on the front end of the motor by the nose-plate, and on the back end of the frame, on which are cast the iron boxes and arms which serve as a support for the armature shaft bearings.

The armature is of the Gramme-ring type, which possesses many points of advantage over the drum type of armature. Any coil on the ring armature can be easily rewound without disturbing its fellows, while with the drum the winding must be removed down to the injured coil and rewound new from that point. It is true that removable coils have been used on drum armatures, but not successfully, for it has been found after extensive and most thorough trials that while these coils are removable, they are practically not replaceable and therefore, that there is no particular advantage in their use.

On the new-type motor the brushes are placed in a horizontal fixed position exactly opposite each other and are easy of access for adjustment or examination. There is no sparking under any of the conditions of operating. A glance at the cuts will show that the field spools are protected at the front and back as well as on the top and bottom by the fields and frame, and consequently injury is almost out of the question, yet as an additional safeguard to these and the armature and commutator, a sheet iron pan is employed which closes in entirely the bottom and sides of the motor and extends above the centre of the armature shaft. Its method of attachment is such that it can readily be removed, permitting easy access to the various parts of the motor, for necessary repairs and adjustment.

The pan, or case, as it may be called, can be extended as high as practice may prove to be desirable, and forms a most effective means of fully protecting the apparatus from snow, dirt and water. The desirability of such a device will be readily appreciated by those who have had coils burned out during the recent heavy snow and rain storms. The mechanical details are as follows:

The spur pinion on the armature shaft is of steel $4\frac{1}{2}$ inch face and has fourteen teeth, number 3 pitch. The split gear on the car axle is of cast-iron with the same face and has sixty-seven teeth of the same pitch.

The speed of the armature shaft, relative to that of the car axle is, therefore, nearly 4.8 to 1. The motor is designed to clear the top of the rails four inches, when mounted on 30-inch wheels. The speed of the armature when the car is running at ten miles an hour, will be 538 revolutions per minute, or the speed of the armature is 53.8 turns per minute per car mile per hour.

ELECTRIC RAILWAY MOTORS. 49

FIGURE 44.

The gears are entirely inclosed in a dust and oil-tight case, which is provided with a hand-hole closed by a spring cover, permitting ready examination of gears and introduction of gears and oil to the interior of the case. The advantage of a single reduction motor having its gears and pinions run in oil and fully protected, over double reduction motors, with two pairs of gears and pinions, unprotected from dust and dirt, is quite apparent. In this new-type motor the first cost and subsequent maintenance of the armature pinion and intermediate gear is surely eliminated. The single reduction gear corresponds to the old intermediate pinion and axle gear wheel, but it has the great advantage of being boxed and run in oil so that the important item of gears will be hereafter practically stricken out of the repair bill. The cost of repairing armatures will be substantially reduced by the new arrangement, but to what extent cannot yet be definitely stated.

Aside from the facility of removing and replacing the armature and the case with which one of our sections of a Gramme-ring can be removed, the slow speed will materially lessen the bursting of binding wires, the displacement of coils and breaking of commutator connections, together with other injuries occasioned by centrifugal action. Considerable noise has also hitherto been occasioned by the brushes bearing on the high speed commutator. This noise, as well as that of the gears, has been entirely eliminated by the new form of machine, so that the rolling of the wheels on the rails is the only sound to be heard in a car equipped with this motor, even at its highest speed.

Another consideration of great value is the form of the magnetic field. In previous Thomson-Houston types of railway motors and in the Edison and Westinghouse motors, the magnetic field is short-circuited outside of the armature, more or less by the support necessary at the nose plate, and some sacrifice of the strength of this nose-plate is required in some styles of eight-wheeled trucks. The effect of this short circuiting on the magnetic field is to weaken the power of the armature. The difficulty is entirely avoided in the new motor which has substantially an iron-clad magnetic field.

The Thomson-Houston W. P. Railway Motor.—*One of the most interesting exhibits at the Pittsburg Street Railway Convention was a new slow-speed railway motor of the Thomson-Houston company, of which the accompanying illustrations give an excellent idea: It has been in process of evolution for six months or more and has been worked up under the careful superintendence of Mr. Walter Knight. The new machine embodies some decidedly novel features and its excellent performance on the special car equipped with it was very favorably commented upon. It is known to the trade as the

FIGURE 45.

W. P. motor, which being interpreted means water-proof, and it well deserves the name, because of the particularly complete iron-clad character of the field magnets.

Fig. 45 gives a perspective view of the motor, and from it the arrangement of the iron is at once obvious. Singularly enough, it is a two-pole machine so arranged on the theory that the comparatively slight gain in weight efficiency that could be obtained with a multi-

* Electrical World.

polar type is more than offset by the increased complication of the windings. The only portions of the machine open to the outside air are exposed at the two oval openings at the ends of the armature shaft, and even these can be easily fitted with covers should such a course prove desirable. The whole magnetic circuit is composed of two castings bolted together and free to swing apart by a hinge allowing ready access to the armature.

Fig. 46 gives an excellent idea of the internal arrangements. The armature itself is very nearly twenty inches in diameter, a very powerful Pacinotti-ring nearly six inches on the face and of about the

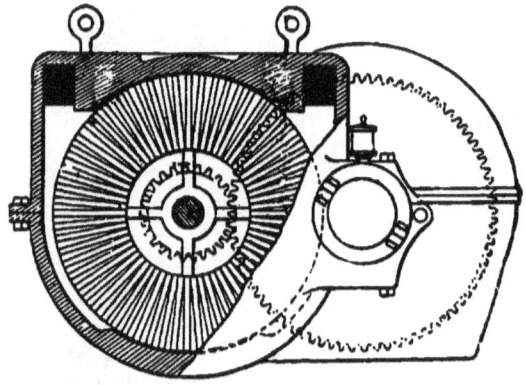

FIGURE 46.

same depth. It is wound with comparatively coarse wire in sixty-four sections, with fourteen turns to the section. Each coil is tightly placed in the space between two of the projecting teeth, and about the interior space the separate coils are closely packed, leaving only sufficient room for the four-armed driving spider.

As will be seen, the armature takes up most of the full height of the machine, the pole pieces being but trifling projections and the requisite cross-section of iron being obtained by extending the poles to form a closely fitting iron box that appears in the exterior view.

An unusual feature is the use of but a single magnetizing coil wound not directly about the upper pole piece but on the casing immediately surrounding it. The lower pole is but slightly raised and both pole pieces are of the greatest extent permissible with the dimensions of the machine. The use of a single magnetizing coil produces naturally an unbalanced field and a strong upward pull on the armature tending to relieve the pressure on the bearings. The iron-clad form, however, tends to distribute the lines of force so as to avoid the sparking and change of lead that might otherwise have to be feared. The single coil is wound with quite coarse wire and its position insures the maximum magnetic effect from the current.

The speed of the new motor is about the same as that of the older S. R. G form, and its general working efficiency is somewhat better, owing not so much to a greater maximum of efficiency as to a better working curve—at both heavy and light loads. The brush holders are shown in the cut, and the slots in which they fit render their position evident. The brushes are of the ordinary carbon description and are readily accessible through the opening at the end of the shaft.

In operation the W. P. motor has been highly satisfactory. It runs with but trifling sparking and no heating to speak of, gives a very powerful torque, and is singularly free from liability to damage of the armature, for which its careful insulation and the Pacinotti form adopted are, responsible. It is now being regularly manufactured at the Thomson-Houston works, and it is expected to take with great advantage the place in popular favor of the S. R. G. motor that has made so good a reputation for itself during the past summer. It is an interesting departure, both electrically and mechanically, and aside from its special features its general qualities of iron-clad field, gears running in oil, and the ease of access to the working parts will commend it to the practical street railway man.

The Westinghouse Four-Pole Single-Reduction Street Railway Motor.—A view of the motor is shown in Fig. 47, bringing out more

FIGURE 47.

prominently the gear casing. The construction of the motor can be readily comprehended by referring to the view. Here are shown the castings complete of the motor, consisting of only three parts — the frame and the two semi-cylinders, the two latter being practically one. The size of the frame is such that it can be placed upon a bogie-truck, being equally well adapted for an eight-wheel as for a four-wheel car. The width of the motor is such that it can be used on a 3' 6" guage. In the sides of the two semi-cylinders are seen the holes where the plates are secured, which serve as a protection to the sides of the machine.

If necessary, the machine can be entirely shut in. It was formerly believed that a motor could not be thus enclosed, since it needed ventilation; but experience with slow speed motors has demonstrated that if a motor be correctly designed, electrically and mechanically, and properly constructed there is no difficulty whatever in enclosing it. At the same time, if, in some cases, it be deemed advisable to allow a small opening for ventilation, the plates can be constructed accordingly.

This method of enclosing the motor is exceedingly convenient in snow and rain storms, and especially where the cars pass over trestles which expose the motor. Heretofore, considerable trouble has been experienced from water dripping on the motor through the car floor. In this motor, as is obvious, such troubles are eliminated. Again, the objections to the motor being exposed to water, dirt and dust can be appreciated when it is remembered that a large number of engineers favor some method of mounting the motors on the car floor. The above objections are overcome by making the motor iron-clad. By again referring to the view, there will be seen the four internal poles; hence, it is called a four-pole motor. Some of the advantages of a four-pole motor over a two-pole machine, are: slower speed; greater simplicity; more symmetrical; and a greater radiating surface for the field coils. In case a two-pole motor is used, and the same amount of wire is wound about these two poles, the radiating surface is far less than where there are four poles.

Another important feature to be noticed is the form of the motor proper; namely, circular. It is a well-known law in mechanics that the strongest form is the arch; consequently, by this cylindrical form, we obtain the maximum strength with the minimum amount of

FIGURE 49.

material. All corners and sharp edges which mean unnecessary weight, and at the same time having a tendency to reduce the efficiency, are eliminated from this machine. The fields are enclosed and protected, not merely externally by the surrounding cylindrical shell, but also internally by a heavy brass cap. There is no liability to accident in case they strike any obstruction in the road, neither can they be injured by gross carelessness in handling.

The cast-iron frame, on which the motor is mounted, forms a distinguishing feature of the Westinghouse machine. This frame is rectangular, in one casting, and made strong at points subjected to the greatest strains. Special machinery has been devised for boring out the holes for the bushings, so that the frame, and, in fact, all parts of the motor, are interchangeable. By means of this frame the armature shaft and car axle are maintained in alignment, and consequently perfect meshing of the gears is obtained, which experience has proved to be of importance. The gearing is mounted closely to the frame, so as to avoid the objectionable buckling and tendency to loosen the moving parts. This method gives a strong

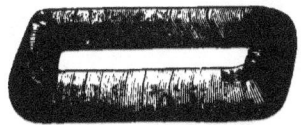

FIGURE 50.

mounting and perfect rigidity between the parts of the motor. Moreover, by extending this frame around the motor and suspending it at both corners, we distribute the strains and prevent the abnormal wearing of the bearings, so characteristic of centre suspension.

The illustration (see Fig. 49), shows the method of hinging the field castings. These, as will be noticed, can be swung back, giving easy access to the fields and armature. It will be observed that the poles protrude radially from the interior of the cylindrical shell. The field coils, one of which is shown in Fig. 50, are slipped over these poles, held in position and at the same time protected from the interior by a brass cap. The ease with which the fields can be removed or replaced needs but a glance to be understood. Any field can be removed without disturbing any other part of the motor, and this can be accomplished in little time. The lower fields can be similarly changed by swinging back the lower semi-cylinder. The armature is then ready to be taken out, and by taking off the brushing cap and placing a sling about the armature, it can be lowered into the pit without obstruction or danger of injuring the same.

The armature is what is known as the drum type, which experience has demonstrated to be superior to other types for street railway work. The armature core is built up of laminated grooved iron plates, so that the completed core has slots to receive the wires. In the armature the wires are imbedded in iron, hence they cannot be injured from ordinary external causes. Since the surface of the armature is iron, the air space, that is, the distance between the iron

FIGURE 51.

of the armature and pole-pieces, is reduced to a minimum, increasing the efficiency of the motor.

The armature shaft is manufactured from the best grade of forged steel, especially prepared for this purpose. The construction of the shaft and armature make it exceedingly strong, and capable of withstanding the severe strains sometimes brought upon it. In looking at the frame, it will be noticed that the oil receptacles are sunk into the same. These oil receptacles are so placed that there is no possibility of injuring them. It is worthy of attention that these facilities for oiling are excellent. The oil boxes are large, and the method of oiling is the same as that of the high speed motor, with which they

ELECTRIC RAILWAY MOTORS.

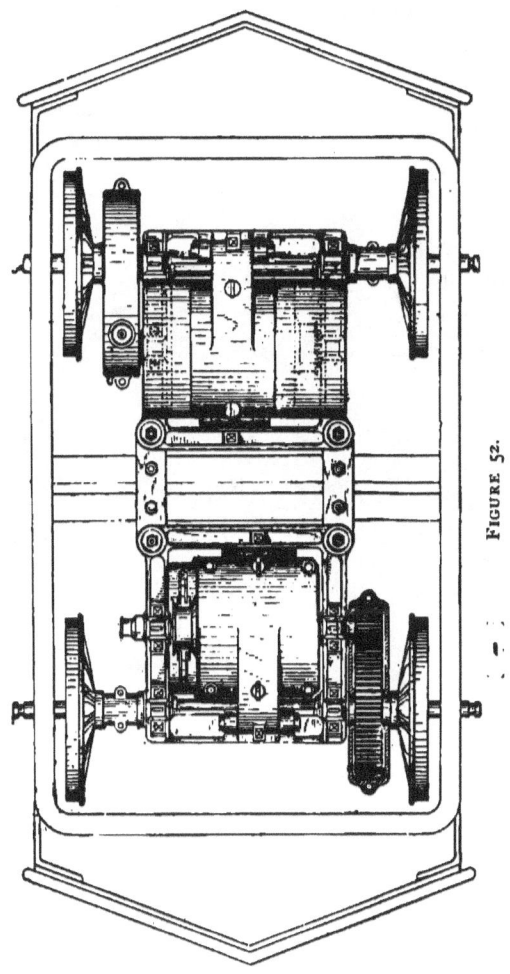

FIGURE 52.

have never had a hot box, so that it can be said with confidence no trouble will be experienced from this source with their slow speed motor.

The field coils are wound with wire having exceedingly large carrying capacity. The arrangement adopted for the brush-holder—see Fig. 51—has also been carefully worked out. It consists of a

FIGURE 53.

square oak holder attached to the side of the frame, and carrying the brush-holders proper, which are clamped so that they can readily be adjusted. The carbon brushes are placed in a sliding frame, and pressed against the commutator by a pair of springs, which can be released by a pressure of the finger, and the carbon slipped out for replacement when worn. The casting supporting the brush-holder is fastened to the bottom of the motor frame, so that the brushes rest on the upper part of the commutator, the greater part of which is exposed above, so that the commutator can be cleaned from the inside as well as from the outside of the car. Fig. 52 shows plan of motor upon truck.

The Edison Slow-Speed Single-Reduction Motors are shown in Figs. 53 and 54. The Edison General Electric Company now build these motors in the following sizes: 15, 20, 25 and 30 h. p.

Only two of the four poles, namely, those in the horizontal plane, are wound with coils, the two in the verticle plane being magnetized by induction from the same spools, and forming, as it were, conse-

FIGURE 54.

quent poles of opposite polarity. The entire field is of special soft-cast steel, with the pole pieces attached by screw bolts after the coils, wounds on vulcabeston spools, have been slipped on over the straight cores. As a result of this construction and the employment of cast-steel, the magnetizing force required is small.

In the construction of the machine the armature bearing and the cylindrical armature space are bored out at one operation, making the armature run perfectly true.

The armature is a Gramme ring with "Paccinotti" teeth. The core is built up of soft punched iron rings, with the end plates of wrought iron, and bevelled. On the interior diameter of the hollow cylinder built up in this manner, there are four grooves placed 90 degrees from each other, and into these grooves the aluminum bronze spiders are pressed by hydraulic pressure, two spiders being employed and bolted together. In this way there is a firm mechanical connection between the armature shaft and the ring, making an armature of extraordinary strength and durability.

The winding, in 140 sections, is put on in one continuous length of wire, and at each section, a tap-wire of German silver is taken to the commutator, the coils being substantially insulated with mica.

The entire machine is encased in an iron cover, and the total weight of the 25 h. p. with gears is about 2,200 pounds. It is intended to apply two of these motors to each car of the double truck type, while one machine alone will be able to drive the smaller cars. The motor is of the series type, and the regulation is effected by varying the combination of the field section.

Carbon brushes are employed, and the brush holders being rigidly fastened to the frame, require no shifting throughout the range of load to which the motor is subjected.

These motors are now fitted with iron and copper gear covers. (See Fig. 54.) These covers completely enclose the pinion and gear wheel so that they may be protected when running, from stones and other pieces of material getting into them.

The copper covers are considerably lighter than the iron ones. The weight of the copper cover complete is 44 pounds and the iron cover weighs 130 pounds. The cover is made in two halves, each half being provided with a flanged edge so that the halves can be bolted together very securely. The cover is held in position on the motor at three points, by means of two angle pieces which are fastened on the cover and bolted to the motor frame upon the axle bracket, and a lug piece on the top half of the cover through which a bolt can be passed into a boss on the armature bearing. The angle pieces which are attached to the axle bracket, also act as locking nuts for the bolts that fasten the axle bracket to the motor frame. To put on a gear cover, take the bottom half and fasten it up under the gears by the bolt through the lug piece, see that it is clear of the gears everywhere and fasten the bottom angle piece on to the axle bracket, then bolt this angle piece to the cover. Having the bottom half in position, place the top half on it and bolt the two together by means of the bolts around the flanges, then put the remaining angle piece on the axle bracket, and bolt it to the top half of the cover. In some motors the lug piece is on the top of the gear cover, in which case the top half is put on first. The covers are fitted with a babbitted bearing for the axle shaft and provided with an oil cup.

The Short Gearless Motor.—The gearless motor (designed by the Short Company) is shown in Fig. 55. Referring to the machine in a general way, it is seen that all gearing is absolutely eliminated, the number of bearings is reduced to two on each motor, and four in the equipment. The armature speed comes down to the minimum, namely, that of the car axles in practical operation. The noise of gearing and the brushes is entirely obviated, and there are but three wearing parts on each motor. The intensity of the magnetic field is now at its maximum; this effect being due, not to a material increase in the weight of armature and pole pieces, but to the wholly different method of construction. Instead of two magnets, we find eight; instead of a wide magnetic gap, we find one extremely narrow, with consequent great intensity of the "field of force." Instead of a drum armature of small diameter, we find a ring armature of comparatively large diameter, and increased "leverage;" the sum total being that we have here in full measure a motor of the second type, namely, one with an armature revolving at low speed in an intense "magnetic field," exerting a power fully equal to the motor with gearing, and at a considerable less expenditure of current, since all friction of gearing is eliminated.

The motor is complete in itself. It is not keyed to the car axle, *nor does it touch it at any point.* The motor as a whole can be taken off the car axle after removing a wheel, but in practice it will rarely or never be found necessary to do this. A plan of the 15 horse-power gearless motor is shown in Fig. 56. A sectional view is shown in Fig. 57.

The field magnets are eight in number, four on each side of the armature. They face each other at a distance of only ten inches and thus form a most intense magnetic field. The magnets are bolted to the framework of the motor, in the center of which are the

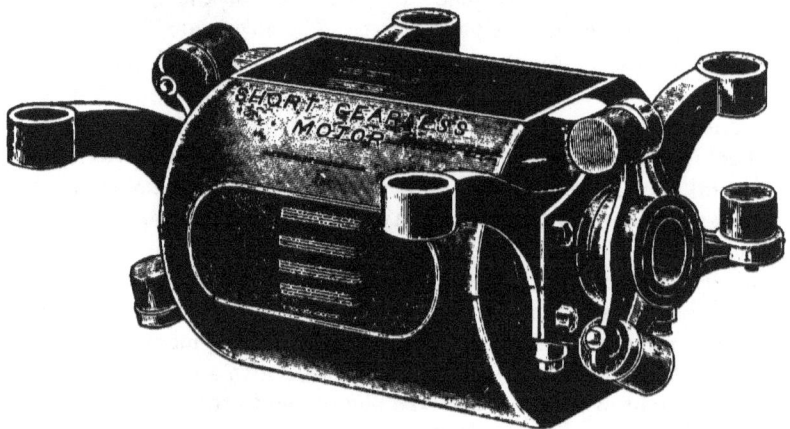

FIGURE 55.

bearings which carry the hollow armature shaft. (See Fig. 58.) The double arms running out from the framework to the cross girders on the truck make provision for the support of the entire motor. The insulation between these brackets and the girders is provided by means of heavy rubber bushings through which pass the bolts. By removing the bolts attaching the fields to the supporting framework, the coils may be quickly taken out, either for repair or to more easily get at the armature.

The armature is keyed to a hollow steel shaft, which is concentric with the axle of the truck, an inside clearance of one inch all around being provided for. The armature proper consists of a laminated iron core upon which are mounted separate and entirely independent coils, following the well-known methods of the Short double reduction type of motor. These coils are perfectly ventilated, and in past practice almost no trouble has been experienced from burnouts. It

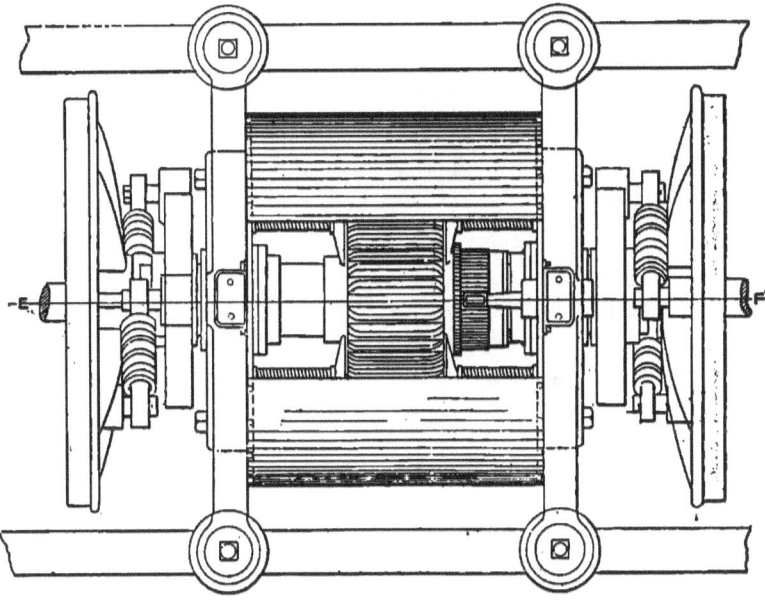

FIGURE 56.

is the one street car armature at present constructed of which it can be truly said that the coils are *absolutely independent*, and can be separately rewound in case of accident, at almost nominal expense. Mounted upon the hollow shaft, close to the armature, is the commutator, which is protected from injury by the surrounding pole pieces. The commutator is massive in construction and of large diameter, the idea being that, because of its massiveness and slow speed, the wear will be reduced to a minimum, and the replacing of the commutator will occur only at long intervals. On the ends of the hollow shaft are mounted two discs fastened thereto, the peripheries of which are insulated from the hubs by the special wooden web construction. Between the commutator and the disc on the one side and the armature and the second disc on the other, are the bearings, which are carried by the motor frame.

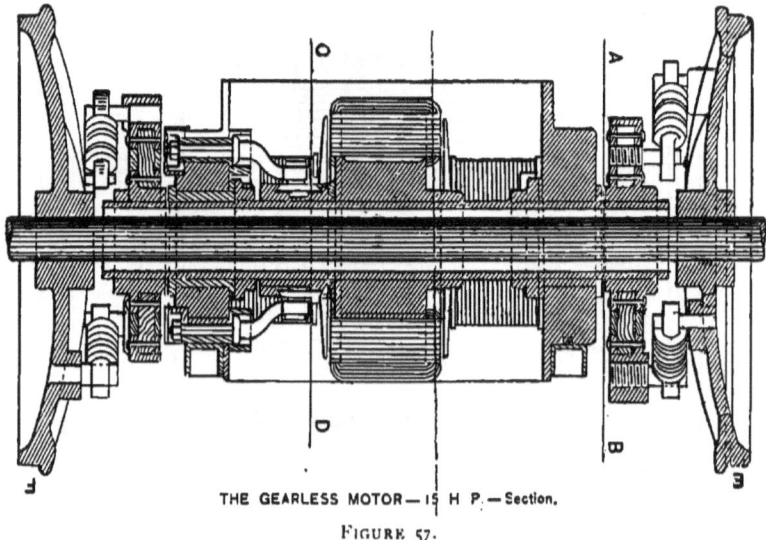

THE GEARLESS MOTOR — 15 H. P. — Section.

FIGURE 57.

It has been before said that the motor has no connection whatever with the car axles; it follows, therefore, that it is necessary to provide means of propelling the car by making some attachment between the hollow armature shaft and the wheels. This is done very simply by means of heavy coiled springs, which extend from the peripheries of the armature shaft discs to bosses on the wheels. Position and attachment of these springs are shown in Fig. 59. They are of great strength, and can pull a very heavy weight with but slight extension or compression. As they are attached to both disc and wheel upon circles of the same radius, their effort is a nearly direct circumferential pull.

From the description above, it is at once apparent that the entire motor is absolutely insulated from the truck at every point. This is a feature which we believe to be of great importance. By this means leakage or accidental connection between field or armature circuits and the iron frame work (which may be caused by moisture, dust,

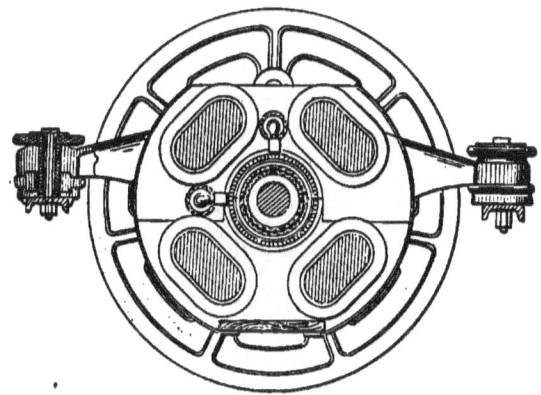

FIGURE 58.

dirt, etc.), does not produce a "ground circuit," and consequent burn-out of field or armature coil, as is the case with other types of machines.

To protect the motor from dust, moisture, etc., which have been a potent source of trouble in other forms of equipment, an iron case completely incloses the motor, except at the top, where necessary ventilation is provided, and is water-tight up to the axles. To get at the motor, it is only necessary to unlatch one end of the casing and swing it down and out away from the mechanism.

The dimensions of the motor are as follows: From the centre of the axle to the bottom of the casing is 12¾ inches. On a 36-inch wheel, which we strongly advise, not only in the gearless, but in other types of motor, there is thus a clearance of 5¼ inches, which is ample for all purposes. At a speed of ten miles an hour, the armature revolves at 94 revolutions per minute, with a 36-inch wheel. The equivalent speed of the single reduction motor would be at least 400, and of a double reduction motor about 1,200. One of the most valuable features of the machine is the facility with which it can be

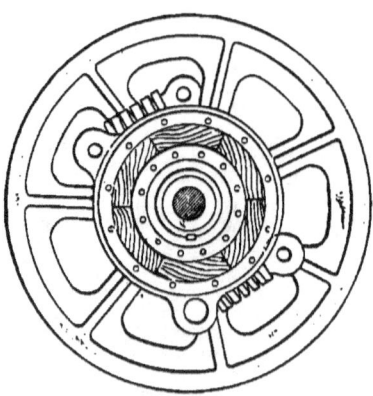

FIGURE 59.

repaired in case of necessity. By loosening four bolts in the motor frame-work, and by taking off the iron strips below the wheel-boxes, one end of the car may be jacked up, and the axle-wheels and armature complete run out from under, into the light of day. The armature coils may be rewound without removing the armature from the car axle. Field coils can be repaired as easily. The commutator may be reached and dressed while the machine is running. If steel tired wheels are used, by a special arrangement the motor may be jacked up, raising the wheels from the ground, current brought to the motor, and the wheels turned just as would be the case on the truck, so that by a special "tool-jig" the wheels may be turned down as required, thus removing any flat spots or other imperfections. Or the wheels and axles may be turned from outside through the hollow shaft of armature, without the least effect on motor, it being, of course, necessary, however, to remove the spring attachment between the hollow shaft discs and the wheels. In case it is found necessary to replace a commutator, a wheel must be pressed off, and the commutator removed bodily. This could be done only with great difficulty if the armature were keyed direct to the axle instead of being on the hollow shaft. The commutator will have a life three or four times that of the wheels in common use on electric railways, and it will not usually be necessary to press off a wheel for the express purpose of replacing a commutator.

CHAPTER V.

RHEOSTATS.

A RHEOSTAT is an apparatus for throwing a variable resistance into a circuit; thus regulating the amount of current to meet the requirements of each case.

Hence, to regulate the field magnetism of shunt and compound dynamos, a rheostat consisting of long coils of comparatively fine wire is connected in the shunt circuit. The current in such cases is only from 1 to 3 amperes. When a shunt, or a series motor on a constant potential circuit, like a railroad motor, is started, a rheostat of shorter coils of comparatively large wire must be put in the armature or main circuit. The reason is that the armature resistance must be low in order that the motor be efficient. With the motor at rest, if the only resistance to the flow of the current were that of the wire on the motor, too easy a path for the current would be open. The motor would start with a tremendous jump that would throw off belts, strip the teeth from gears, besides endangering the wire to melting. A rheostat should offer sufficient resistance so that the current flowing will at no time be beyond the safe carrying capacity (in amperes) of the motor armature wire. As the motor turns, it generates a counter electro-motive force which pushes back on the primary electro-motive force with which the motor is supplied. This counter electro-motive force cuts down the current similar to the action of a rheostat. As the speed of the motor increases, the rheostat resistance can be gradually withdrawn. It will be easily seen that a rheostat should not be kept in a main circuit for any length of time, as by so doing the motor will receive a less potential than necessary, and would work at a reduced efficiency.

Rheostats get heated from the flow of the current, showing that useless energy is being consumed.

They are a necessary part of the equipment of a power switchboard. In such cases the main current from the dynamos does not go through the coils of the rheostats, but only that portion flowing through the shunt field winding. It is impossible to make two dynamos so exactly alike, that without any regulating devices, they will both generate at the same potential. The varying magnetic qualities of iron, and the different resistances of the wire when cold and heated alone make considerable differences. Pulleys are not turned to exact diameters, and all belts do not have the same adhesion, so a variation in the speed of the dynamos must be allowed for. A rheostat in each field circuit is an easy and effectual device to accomplish a variation of 20 per cent, the total electro-motive force of the dynamos, without interfering essentially with the working efficiency. As the current is small, the coils of the rheostat can be made of small wire, usually German silver, about $\frac{1}{16}''$ diameter. To secure close regulation, the switch to which the resistances are connected is divided into a large number of points: forty, or even eighty different segments are arranged in a circle, and a shoe is arranged to slide over these when a hand-wheel is turned. More or less resistance can be thus put in the circuit and the magnetization of the fields increased or decreased.

Rheostats are usually included in an electric car equipment. The only other method of controlling railroad motors is to use different combinations of connections of the field magnet coils to secure a variable resistance; however in this case, some of the coils need to be of German silver to offer sufficient resistance, and their use is attended with heating, dangerous to the motors. With a rheostat, nearly all disorders of the equipment will show themselves clearly in that one place, and repairs can be easily and cheaply made. With the other, known as the "controller" system, the motors themselves have to suffer, and repairs are inconvenient and more expensive. It is a strong point, however, with such a system is that the current is more economically used in starting the motors.

72   ELECTRIC RAILWAY ENGINEERING.

The Thomson-Houston Rheostat (see Fig. 60) has the resistance laid closely in a semi-circular iron trough. This trough is connected by radial spokes with a center or hub that supports a vertical insulated steel spindle. On this spindle turns a wheel or drum, actuated by the driver through means of a sprocket chain and steel cable. The drum moves a brass arm, on the outer end of which is a shoe that rubs on top the resistances in the trough.

The resistances are made in the form of thin iron punchings, of a

FIGURE 60.

shape shown in Fig. 61. Although the piece is only about $2\frac{1}{2}'' \times 4''$ the path through it for the current is long. The current enters one end, at "A," and is obliged to travel a crooked path to arrive at "E." Another punching exactly the same touches the parts "E," of each, together, being separated from the rest of the surface by a thin sheet of mica. In this second punching the current travels from "E" to "A"; thence it passes by contact into the part like "A" of the third punching. A sheet of mica keeps the punchings from touching except at the extreme ends. So on, in the series, the current goes up and down, back and forth through the sheet iron until the requisite amount of resistance has been reached. At regular intervals, thin cast-iron plates, with thickened upper edges, are inserted. These stand about half an inch above the punchings, and are the parts that receive the contact with the shoe on the movable arm. These contact plates are thicker at the outside edge than at the inside, and thus compensate for the difference in length of the inner and outer semi-circumference of the trough.

The bottom and sides of the trough are lined with mica and slate;

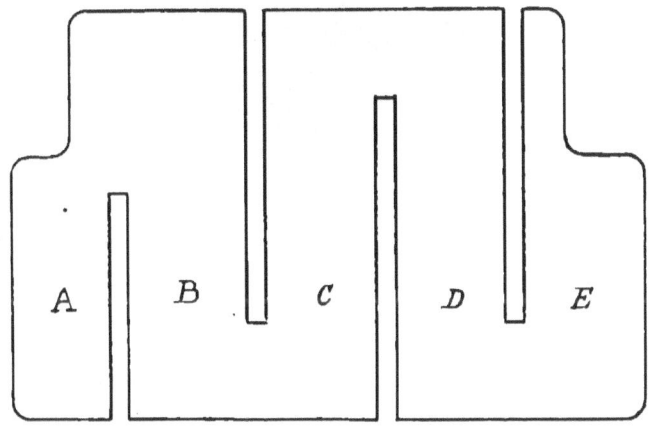

FIGURE 61.

the plates are held down in place by two semi-circular iron bands pressing (with mica insulation) on the shoulders at "A" and "E." So it is seen that the entire construction is iron, mica and slate—articles either cheap or incombustible. These rheostats can be kept heated red hot for hours without appreciable damage. The resistance is about 20 ohms and the capacity 60 amperes.

Stops for limiting the extent to which the arm can be moved are provided, and at the place where the contact is made and broken a magnetic "blow-out" is located, which quickly extinguishes the arc that follows the breaking of the circuit.

The Short Electric Company's Rheostat (see Fig. 62) is, in some respects like the Thomson-Houston. Sheet-iron plates, slotted to produce a similarly long path for the current, are used to give the requisite resistance. Thin sheets of asbestos separate the plates, except at the ends, where one touches its neighbor to keep the circuit complete. Instead of being contained in a trough, these laminations of iron and

FIGURE 62.

asbestos are strung on iron bolts; a bushing of lava, however, keeps the plates from actual contact with the bolts. This construction admits of air spaces being left around the different groups or sections of which the entire rheostat is built. The whole piece of apparatus is suspended from a wooden frame under the car body. "Leads" of copper wire are taken out from the plates at regular intervals, and are carried to corresponding points of a switch. This switch is attached directly under one of the platforms of the car and the arm that makes contact is operated from a crank the same as in other systems.

CHAPTER VI.

ELECTRIC HEATERS.

A NEW feature in the application of electricity is its use for heating purposes and its practicability is now receiving attention. Its principal merits are cleanliness and convenience, although at present it is an open question as to its economy. We have no doubt but that within a short time it will be so perfected that economy will be one of its strongest claims. The Burton Electric Company, the original manufacturers, control absolutely the patents of Dr. W. Leigh Burton, covering the simplest and most effective devices for electric heating.

The heaters require for constant running 3 amperes on a 500-volt circuit, which is the customary current for street railway work.

3×500 volts $= 1500$ watts $\div 746 = 2$ h. p., required to keep the car warm under ordinary circumstances.

When first warming the cars, it is found desirable to operate the heaters in multiple for a few minutes. This is rather expensive, as the current required to operate the heaters in multiple is 12 amperes on a 500-volt circuit.

12×500 volts $= 6000$ watts $\div 746 =$ a little over 8 h. p.

But taking into consideration hauling coal, care of fire, dirt, etc., the electric heaters have a strong claim for use in street car service.

The Burton Electric Car Heater is shown in Fig. 63. In outward appearance the heater, as constructed for street car use, is a flat, corrugated iron casting twenty-seven inches long by eight inches wide, mounted upon iron legs, which raise it four inches from the car floor.

It consists of two corrugated iron castings, holding in the intervening space the resistance wire, imbedded in finely powdered, dry fire clay, the purpose of which is to readily absorb the heat as generated in the wire, and prevent the oxidation of the latter. The corrugation of this receptacle is unusually pronounced, in order that the radiating surface may be as great as possible. All joints are thoroughly made, thus preventing the slightest loss of fire clay. Four heaters comprise the equipment for an ordinary street car. As shown in Fig. 64, the heaters are placed two at each end of the car under opposite seats. The group is so wired that with the combination switch furnished with each set they may be operated either in parallel series of two or in direct series of four.

Provision for such an alteration in circuits is made in order that a

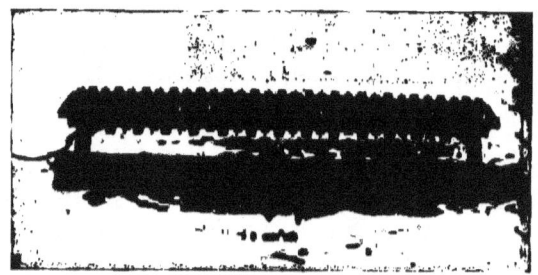

FIGURE 63.

car may be heated rapidly before starting on a trip, while thereafter a continuous current of lesser amperage sustains the acquired temperature. No. 10 or No. 12 triple braid weather-proof wire should be used to connect the heaters of a set. Leaving the main circuit on the line side of the motor switch, it should pass from heater to heater and make its ground connection at the regular ground binding post on the car truck.

Best results, moreover, are obtained when the heaters are set within tin cases, so constructed as to reflect the heat into the car and to prevent, as far as possible, any distribution beneath the car

ELECTRIC HEATERS.

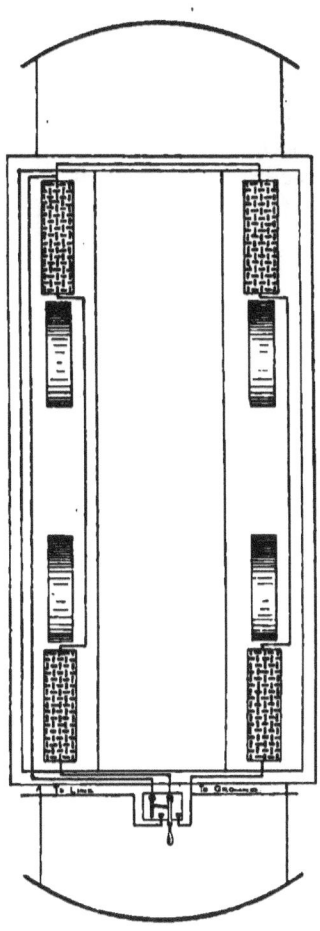

FIGURE 64.

seats. In cases where the seats are paneled, the panel directly in front of each heater should be removed and the reflectors so arranged that all the heat from the heaters will be reflected through their respective openings. While not necessary, an improved appearance is obtained by having screens fitted to the openings. It should be understood that the heater once set up may remain in the car the year round, without the slightest inconvenience to employes of a road or its patrons. They occupy space otherwise unused and their presence never reduces the seating capacity of the car. The fact that the heaters once set up require no further attention justifies the small trouble and expense of first-class arrangement.

CHAPTER VII.

TROLLEYS.

THE trolley apparatus is used to make the contact with the overhead wire and the current is passed through it to the motor upon the cars. It usually consists of a small brass grooved wheel five or six inches in diameter centered in graphite or raw hide bearings and mounted upon the end of a pole ten to twelve feet long. The pole is pivoted upon a frame which is fastened upon the roof of a car in such a manner that it trails along from the middle of the car. At the lower end of the frame is a spring, or springs, which press the wheel against the trolley wire. By the actions of these springs the contact with the trolley wire is kept unbroken as its height varies. By a cram arrangement the springs are made to act in such a manner as to equalize the pressure of the pole in any position against the wires at the various degrees of expansion or compression of the springs. The object of pivoting the pole is to make it flexible so that it may move in any direction. The pole may be made of wood (sometimes bamboo) or of steel. Steel poles are made in several forms; they may be straight, tubular, drawn tapering in one piece or in sections of different diameter. In some cases the current is made to pass through the journal of the trolley wheel or through brushes and is carried through the pole which is insulated at its base to the wire leading to the motor.

Figure 65 shows the trolley pole and wheel used by the Rae system. The pole is made of tubular steel, and is drawn to three reductions, being $\frac{1}{2}$ inch in diameter at the top, and one inch at the bottom. A pole of this kind is quite light and strong. Fig. 66 shows a trolley stand of the same system, from which the reader may see the

FIGURE 65.

FIGURE 66.

TROLLEYS.

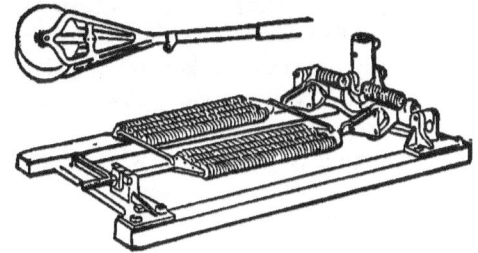

FIGURE 67.

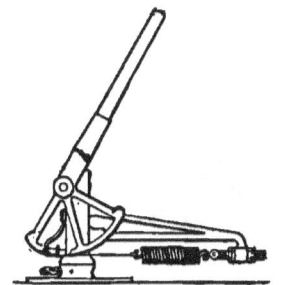

FIGURE 68.

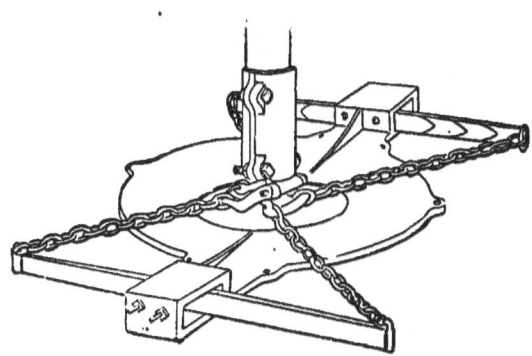

FIGURE 69.

arrangement of the springs and leverage which equalize the tension. Fig. 67 shows the Boston trolley, which is used by the Edison and Westinghouse systems. Fig. 68 shows Baker trolley-pole and stand; Fig. 69 shows Common Sense trolley base; Fig. 70 shows Short Sliding trolley. In this case the wheel is replaced by a shoe which slides along the wire. The shoe is usually lined with soft metal, which

FIGURE 70.

is replaced every few days at a slight cost, as it soon wears out. An old form formally used by the Sprague Electric Co., has lately been revived by Siemens and Halske of Berlin. The wheel is replaced by a bar of metal, which gives the form of a T. The object of this form is to do away with overhead frogs, but it has the disadvantage of excessive sparking.

The Wightman Electric Manufacturing Companies' Trolley, has a circular base having teeth on its upper rim to engage with corresponding teeth on a disc shaped flange of the trolley socket. The lifting springs pull, through the medium of a chain upon a projection centered to the disc underneath. As the trolley follows the wire around curves the trolley socket rotates slightly by means of the teeth described which keeps the trolley wheel always parallel with the wire. (See Fig. 71.)

FIGURE 71.

CHAPTER VIII.

LOCOMOTIVES FOR HEAVY TRACTION.

THE continued success of electric street cars and the demands made by street railway companies for larger and more powerful motors to handle their cars, has led others interested in transportation to investigate the advantages of electric locomotion, with the result that not a few electric tramways are in operation hauling freight about in cotton mills, iron works, mines, etc. In this department of work, also, there has been a constant demand for more powerful motors, so that where the electric locomotive formerly hauled one or two cars, it is now required to haul a good-sized train. There are also at the present time under process of manufacture by several of the leading electrical companies fast passenger electric locomotives, which are designed to travel at a higher rate of speed and to surplant the steam locomotive. Recently two of leading companies, the Thomson-Houston Co. and the Edison Electric Co. have placed upon the market electric locomotives for the purpose of heavy traction, such as hauling freight cars, etc.

The New 100 H. P. Thomson-Houston Freight Locomotive, of which two views are given (see Figs. 72 and 73), has a capacity of one hundred horse-power. The Whitinsville Machine Co., of Whitinsville, Mass., for whom the locomotive was built, purpose to carry their merchandise back and forth from the railway station to their works, a distance of 1½ miles, by means of electric power.

The locomotive is built in a square form with a platform for carrying loads, and cow-catchers and draw-bars at each end. The power is to be furnished by a large generator located at the works of the

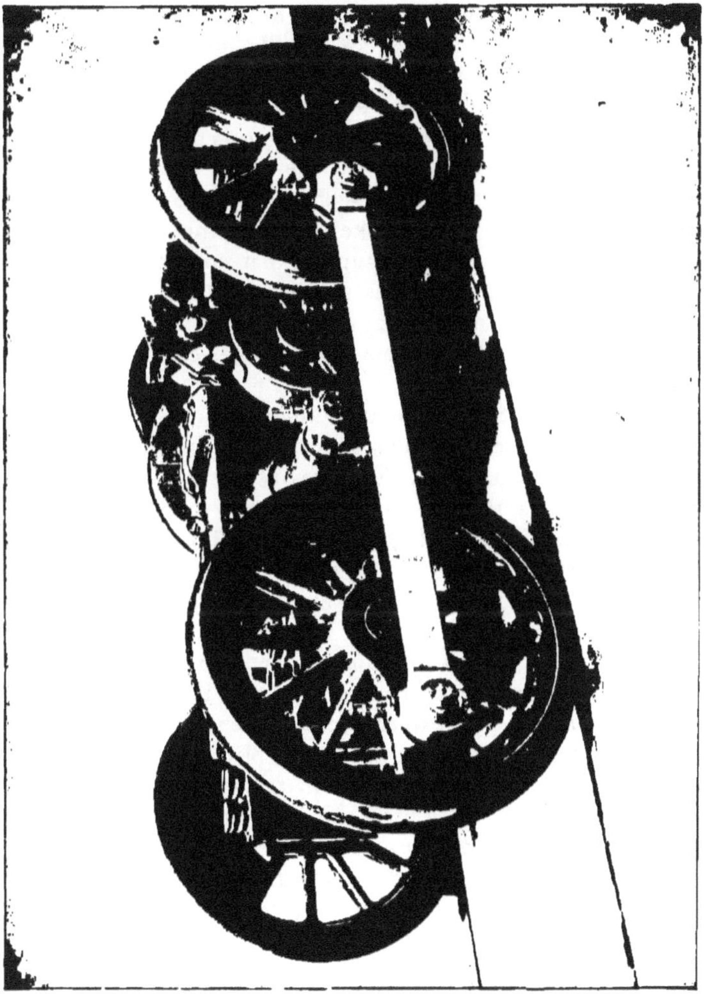

LOCOMOTIVES FOR HEAVY TRACTION.

FIGURE 72.

Whitin Machine Co., and conveyed over a trolley wire from which it is taken by means of a universal trolley bar attached to the locomotive. The construction of the truck is well shown in the engraving. The motor employed is one of the well-known "G" type of the Thomson-Houston Electric Co., and the power is communicated from the armature to the rear axle by means of double reduction gearing, and from the rear axle to the forward one by means of parallel rods. The motor consists of wrought iron field magnets, which are bolted to magnetic yokes of mitis iron. One of these yokes carries the bearings which support that end of the motor on the axle, while the other yoke is spring supported from the other axle. This keeps the gears always in line, and meshing correctly with each other, and at the same time provides considerable spring support for the motor.

The gearing consists of aluminum bronze pinions and mitis iron gear wheels. This gearing runs in gear cases, in which a plentiful supply of grease is placed. This decreases the noise, friction and wear, and increases the life of the gears very materially. On the intermediate shaft is heavily keyed a mitis iron brake drum, which is covered with wood lagging. It is embraced by two half bands of steel, tightened upon it by means of the brake drum lever, situated in the operating stand.

The wheels are 42 inches in diameter, and are heavily steel tired, and the frame consists of two heavy side plates, in which are located the main axle bearings. Two heavy cast-iron end plates in which are cast the cow-catchers, are bolted to the side plates by means of heavy through bolts, which are a driving fit in reamed holes. These end plates carry the heavy spring draw-cars and bumpers.

The operating platform is located at one end of the main platform, and is encased in a railing and covered with a protecting roof. On this platform are located the levers for operating the controlling mechanism, the brake and the double-acting sand boxes. The universal trolley bar also extends upwards from the locomotive at this point.

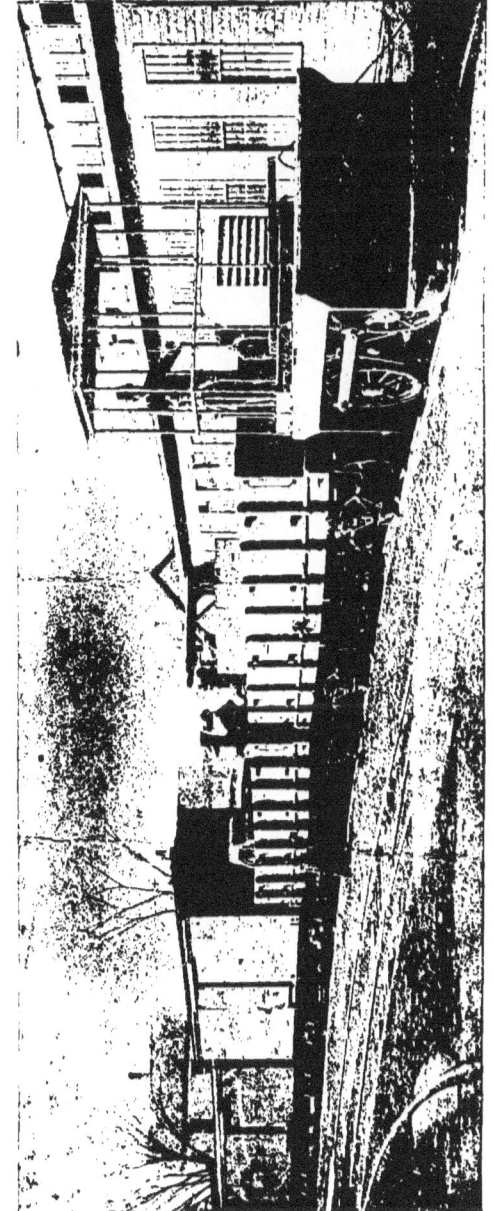

FIGURE 73.

The controlling mechanism consists of two large rheostats of the well-known Thomson-Houston railway type. These are so arranged with their contact shoes that no reversing switch is needed. The operator stands so that he always faces in the direction in which the locomotive is to go, and being in this position he pushes the rheostat lever from him to make the locomotive go forward, and pulls it towards him to make it go backward. A positive centre lock is provided, so that in turning the current off, there is no danger of passing the neutral point on the rheostat, and so reversing the locomotive with the current on. When the operator stands in the above mentioned position, he pushes the brake lever from him in order to apply the brake. The bands are so arranged on the brake-drum that the friction tends to tighten them up more upon the wood lagging, and so assist the operator in braking the train.

The following data give the details of construction of the new locomotive, the construction of which has been under the direct supervision of Mr. J. P. B. Fiske, who is in charge of all the motor work of the company, except that relating to street railways and long distance transmission :

Wheel base	6' 4"
Diameter of wheels	42"
Speed reduction between armature and axle	1 to 25
Gauge	4' 8½" standard
Wheel base	6' 4"
Measured height above rail platform	4' 4"
Greatest length of locomotive (at cow-catcher)	15' 9½"
Greatest length of platform	12' 7¼"
Greatest width of platform	7' 1¼"
Weight of complete locomotive, less trolley pole	42,525 lbs.
Approximate weight of motor	5,400 lbs.

A combined main switch, lightning arrester and fuse-box is placed within easy reach of the motorman, so that he can instantly shut the current off from the locomotive by a slight movement of the hand.

The construction of the motor is of the most rigid and waterproof character, the field spools having their wire enclosed and entirely sewed up in canvas bags, which are covered with a heavy coating of waterproof paint. The locomotive, which weighs 42,525 lbs., is designed to operate at 500 volts. This will enable it to pull a train of four to six heavily loaded cars, or an aggregate load of 200 to 300 tons, at a speed of five miles an hour on a level.

The Edison Electric Locomotive (see Fig. 74) was designed specially for mining work, but is easily applicable to any class of factory employment, whether within the walls, or on the tracks communicating with the main lines of steam railroad. It has a 70 kilo watt motor which is of a multipolar type designed to run on 500 volts with 20 amperes, at 700 revolutions per minute. At this speed, it is calculated that the motor will propel the locomotive and pull a train weighing 110 tons on the level at the rate of seven and a half miles per hour. The armature is of the Gramme-ring type, the commutator having 150 divisions and receiving current from four carbon brushes. The magnet frame is of cast iron and has four inward poles upon which the magnet winding is placed. Projections cast on the magnet frame form supports for the bearings. The power is transmitted from a pinion on the armature shaft, to a gear wheel on an intermediate shaft, a pinion on which meshes into a heavy gun metal gear on one of the wheel shafts. The four wheels are arranged outside of the frame and are connected two and two by connecting rods on either side of the frame. Powerful brakes are provided for each wheel actuated by one brake lever which applies the brake to all the wheels at the same time. The wheel base of this locomotive is 44 inches by 44 inches; 44 inches being an ordinary mine gauge. The motor is controlled by means of three switches, main, reversing and regulating. The main switch is used for starting and stopping the locomotive, the reversing switch for running it in either direction, and the regulating switch for increasing or diminishing speed. The regulating switch is constructed on

FIGURE 74.

the same lines as the regular street car switch, and is used for commutating two fixed resistance coils and the field windings of the motor. The resistance coils are very substantial, being constructed of sheet iron discs; they are fixed at the opposite end to the driver. Sand boxes are supplied for sanding the rails in case of the wheels slipping. These are controlled by two handles arranged on the right of the switch board. The driver has thus all of the six controlling levers within his reach. Draw head boxes are fixed at either end, so that the coupling link can be secured at any desired height. A trolley similar to that used on a street car is fastened on the top of the frame and brings current on to the motor.

The weight of the locomotive complete is about 20,000 lbs., and the overall dimensions are as follows :

Height, 49 inches,
Width, 57   "
Length, 128½ "

CHAPTER IX.

TRUCKS.

THE same logic that was used in the comparison of electric and horse car lines applies equally well to the rolling stock. The cars suitable for horse draught sustain small strains, and the pedestals supporting the axle boxes are usually attached directly to the side timbers of the car bodies and the wheels are light and the axles small. In electric cars the driving power acts as a torsional strain on the axle; this is usually made $3\frac{3}{8}$ inches in diameter, with wheels of corresponding strength. The axle boxes are held in frames that form a truck independent of the car body. The car body can be removed and replaced much the same as in the case with steam cars. The motors hang from the axles for one bearing; their other ends pointing toward each other and connect with a flexible suspension upon arch bars or I beams that cross the centre of the truck, parallel with the axles. As in other departments of manufactures, different firms produce varieties of truck construction.

The Thomson-Houston Company was the first to build special electric car trucks. In these, four wrought iron forgings are bolted together to form a rectangle. The "side bars" are bent into an inverted square, U shape, where each axle box is located, being loose enough to allow for the necessary up and down movement. Two bars of flat iron, one arching upwards, and one downwards cross the car half way between; the end members of the truck afford the "nose" support of the motors. Springs and fastening devices serve to connect the truck frame with the car body. In spite of the double set of springs used, electric cars ride none too smoothly.

The Edison Company are now using a truck of their own manu-

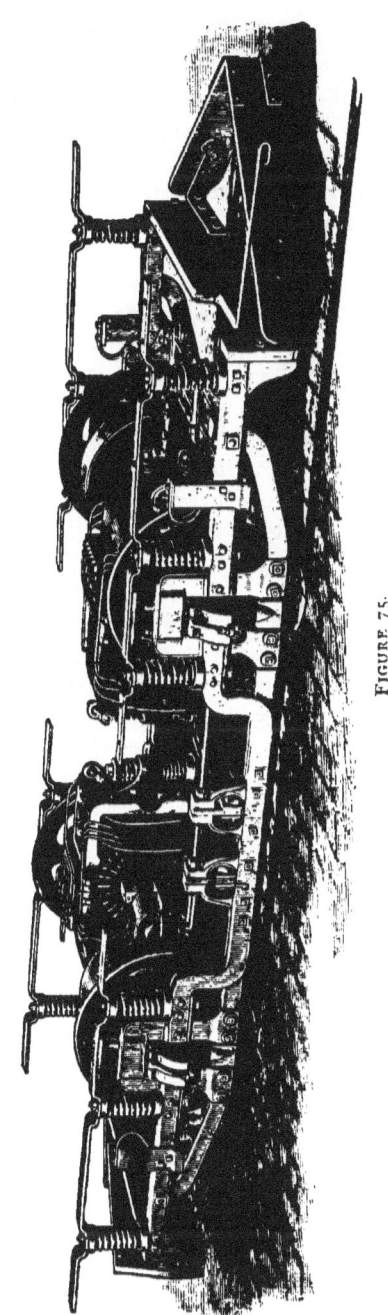

FIGURE 75.

facture in which an illustration is given in Fig. 75—they formerly used the Stephenson truck with most of their equipments; this is unlike the Thomson-Houston, in that the side bars are made of wood and have the pedestals bolted to them, instead of directly to the car bodies. The arrangement of springs is essentially the same as in constructions by other firms.

The Rae Truck (see Figs. 76 and 77) is of quite different construction from either the preceding, due to the different application of the motive power. As but one motor is used on each, the two axles need to be kept in pretty definite relation to each other; otherwise the gearing would be damaged. The construction is a rectangle of I beams firmly riveted together, other cross beams support the motor and the bearings for the bevel gears. Such a car runs well on smooth track, but on uneven roads the strains to which the frame is subjected is severe. Instead of ordinary gun metal bearings for axle boxes the Tripp Car Company are introducing roller bearings with good results.

For long cars the single truck is insufficient. The Robinson Radial Car Company have brought out an ingenious six-wheeled car that has met with excellent success. Each axle is mounted in a separate truck; the end ones swivel on pivots, and turn under the influence of the centre truck as the wheels encounter curves. The car body rests on all three trucks, but the weight is transmitted through rolls, so as to allow easy movement of the mechanism. The end axles are driven by motors, the centre one running idle.

Long cars are also fitted with "bogie" trucks, similar to a steam car. Such a truck has usually two axles, and four wheels and pivoted to the car body, to allow the necessary swing on curves and turns and they make very easy riding cars, as all ordinary shocks are neutralized before the car body has received any jar. Usually two motors are on every car; with "bogie" trucks both motors are at the same end of the car, the other truck running idle. If a motor is put on each truck, the wheels will not take switches so well, and the wiring will be more complicated.

TRUCKS.

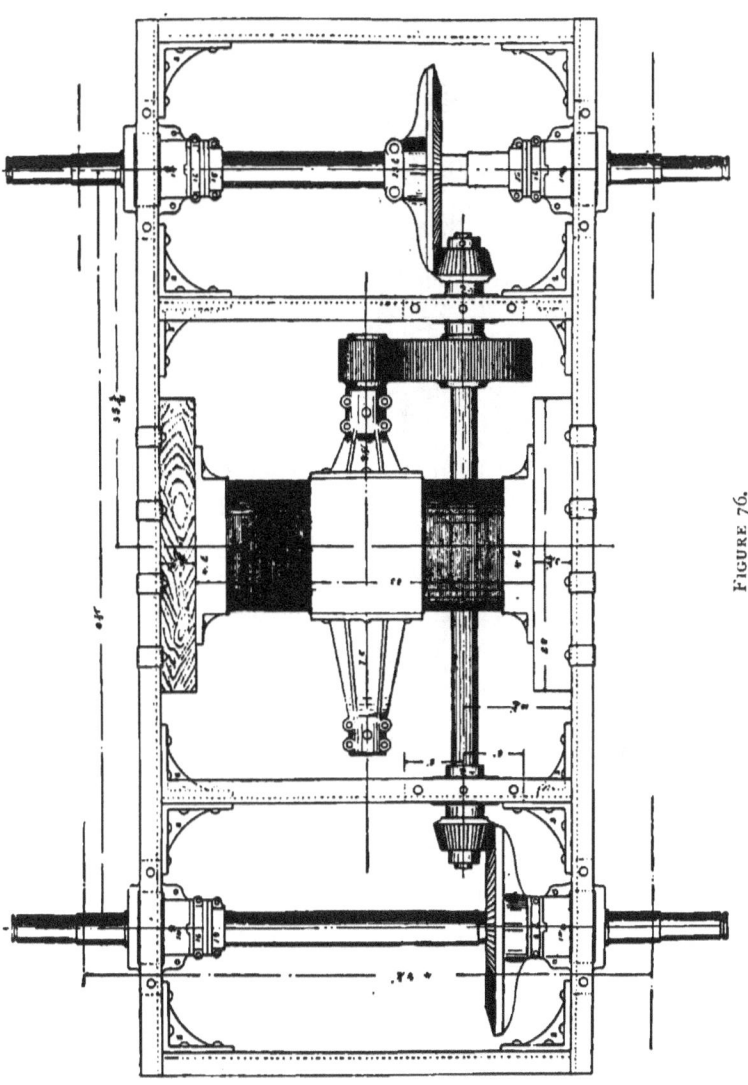

FIGURE 76.

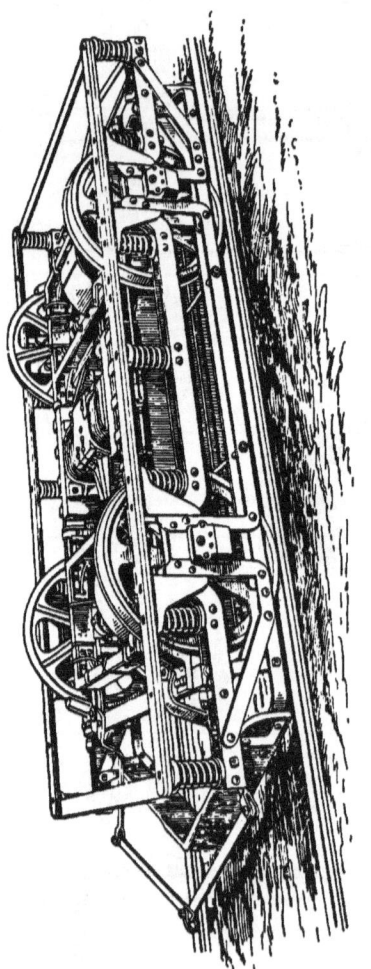

Figure 77.

The Brill Car Company has put out a new truck which is intended for one motor only. In this, the drivers are wheels of larger diameters than the other two, and the pivot is close to the axle of the former consequently the weight of the car comes principally upon those wheels which need the weight for traction purposes.

Many other designs of trucks are made but space will not allow for their description here. Those given above are the principal ones and the author trusts that the reader may obtain a fair understanding of trucks in general from these descriptions.

CHAPTER X.

CAR WIRING.

CAR wiring requires considerable skill. The difficulty of keeping good insulation tires the wireman's patience, while the rattle and shaking of the car keeps the wires in constant danger of loosening from the binding clamps, and of breaking into pieces. Instead of solid wires, cables heavily insulated, such as Clark or Okonite brands are usually employed. No. 6 is a convenient and suitable size and all joints should be carefully wound with tape. When a rheostat is employed the wiring is simplest. Even then there are three cables for each motor for field connections, and the two brush cables. With "controller" systems which use various combinations of the field coils to effect the regulation, even more cables are usually employed.

In wiring, the cables must be left long enough to allow for the jolting of the car body on the truck, yet not long enough to dangle and chafe against the brake rods, or the chains or cables that move the rheostat and reversing switch. It has been usual in previous years to employ but one reversing switch to change the direction of both motors. With the high speed double reduction type of motors, this worked satisfactorily. Single reduction motors have so low electrical resistance in the armature that the difference of pressure or conductivity of the carbon brushes, or the condition of the contact of wheels with rails, will result in sending more current through one motor than through the other. A separate reversing switch for each motor so arranged that the circuits for the two motors are kept separate will balance this resistance so that each motor will take equal amounts of current.

With "controller" systems a separate reversing switch need not be used, but can be incorporated in the mechanism of the "controller stand." With the latter method of regulating street car motors, cables carrying currents under the maximum difference of potentials, are very close—even cross each other. This consequently endangers short circuiting.

Fuses and lightning arresters, necessary adjuncts to an electrical installation, are also in full demand upon electric cars. For lighting the car, incandescent lamps are used. When the pressure of the line current is 500 volts, 100-volt lamps are used. When the pressure of the line current is 550 volts, 110-volt lamps are used. Five lamps are placed in series. Each of the circuits of the car are independent. There is one for each motor when they are in multiple, one for the lightning arrester, one for the lamps and when electric heaters are used, one for them, (see chapter upon electric heaters); of course there are some variations from these rules in different systems. To illustrate, diagrams of two systems of car wiring are given, namely: The Edison and the Thomson-Houston systems.

Fig. 78 shows plan of Edison system. Connections upon motor board as shown in diagram read:

o B+,
o A+,
o C+,
o C—,
o A—,
o B—.

The Cut-Out Switch is the point at which the current divides and goes to the machines (they being in parallel). Its principal object is to allow an easy way to cut out either machine when necessary, each of the main wires running from switch to switch having corresponding wires running from them to machines.

The cut-out as shown in diagram reads: +Arm., +C, +B, +A, —A, —B, —C, —Arm.

The Controlling Switch is the switch from which speed of the car

100   ELECTRIC RAILWAY ENGINEERING.

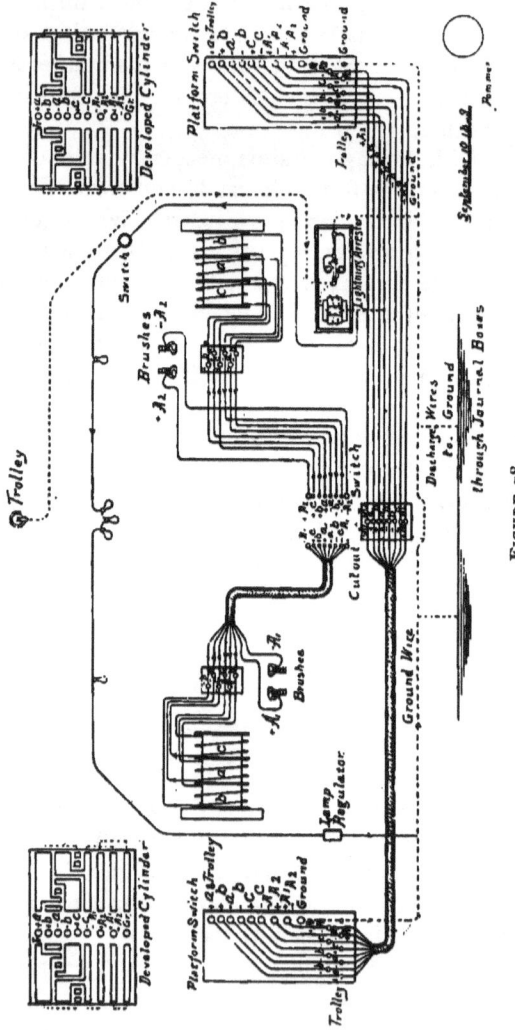

FIGURE 78.

is controlled. The plates on roller are placed so that they either complete a circuit, or break or short circuit it; according to what sections the machine is designed to use on the respective positions. They also place the sections in series or parallel as required.

First Position. The current goes through A coils, then through B and C to armature, to rheostat or slow start, device to ground, always going to switch after passing through each section before going through another. Second Position. In some cases it is the same as the first, excepting that rheostat is cut out, but as shown in diagram the A section is short circuited, or is not used, the current going to B and C, to armature, to ground. Third Position. The A section is cut out, as A minus button rests on a dead plate there is no circuit. The other sections same as second position. Fourth Position. This position places A and B sections in parallel, that is, the current is divided in two paths. The current that goes through A section does not go through B, but goes direct to C, as also does B's current, thus placing C in series with them. Fifth Position. The A and B sections in parallel, C section short circuited. Both of the C buttons rest on the same plate on roller, as also does the A minus and B minus rest on same plate. The current goes from them direct to armature. Sixth Position. The same as fifth. The C minus button resting on dead plate on roller. Seventh Position. The A, B and C sections are in parallel, current going directly from each section to armature. The C minus plates and armature plus plates are connected by a wire in roller, as also are some others, as can be seen by referring to diagram.

A plan of the Thomson-Houston Electric Companies' standard car wiring is given in Figs. 79 and 80. In this plan there are two W. P., R. R. motors in multiple.

    C, C, are the two Railway Motor Switches.
    D, is No. 15 Light Branch Switch.
    E, is the Reversing Switch.
    F, F, are the M. F. S. P.—No. 10—35 Light Cut-out Boxes, (using 4 ampere fuse-wire.)
    G, Railway Motor Cut-out Box.

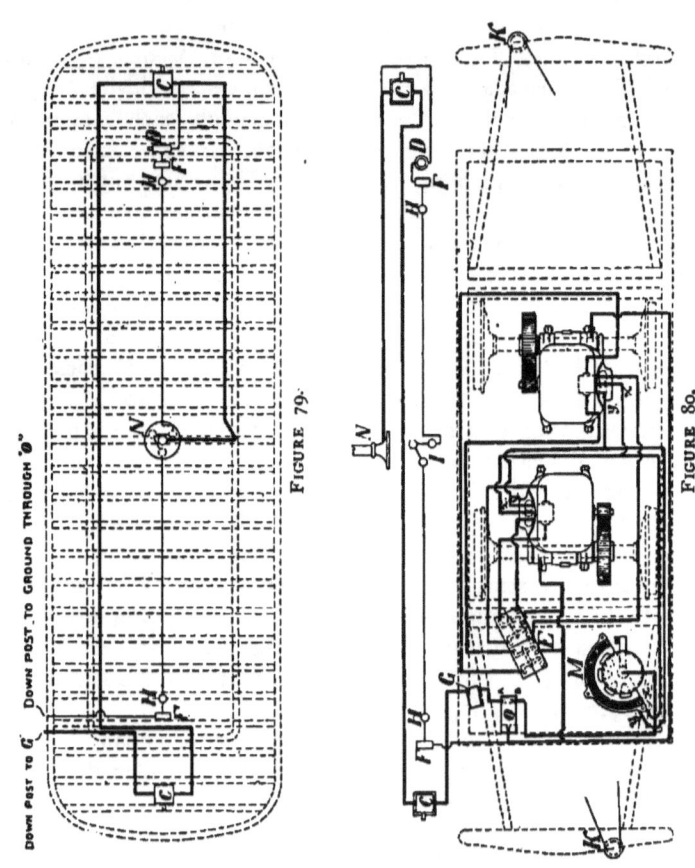

Figure 79.

Figure 80.

H, H, are the Pilot Lamps.
I, is the Cluster of Three Lamps in Car.
K, K, are the Controlling Stands.
O, is the Lightning Arrester.
M, is the Rheostat.
N, is the Trolley Stand.

The path of the current in the car circuit is as follows: From trolley wheel to trolley base to one corner of car to switch C, over motorneer's head, to opposite end of car through a similar switch C; then down on corner of the car to fuse box or "cut-out" G, through lightning arrester O. From here one wire leads to "ground" on both car axles, through which the current flows when a lightning stroke is received. The current regularly passes to the center stud of the rheostat M. When the rheostat arm touches the contacts in the semi-circular trough, the current passes by wire from Y, to each field spool on the motors; then to armatures of each motor, back to reversing switches to "ground." If the rheostat arm is moved over to the extreme limit, the contact is made with a plate connecting with wire X. The current then passes through only part of each field spool; that is, cuts out part of the winding. A reduction in the intensity of the field magnetism results, and allows more current to flow through the armature, giving highest speed of revolutions. When running with these connections the motor is said to be on its rheostatic coil or "loop."

The circuit for the lamps is taken from the main wire near the switch C, thence through the switch D, to fuse-box F, through the five lamps in series H, I and H, through fuse-box F on other end of car through lightning arrester O, to ground.

CHAPTER XI.

THE STORAGE BATTERY SYSTEM.

THE most familiar form of storage battery or accumulator cells consists of lead plates coated with oxides of lead, immersed in dilute sulphuric acid, which after being acted upon by having an electric current sent through them for a certain length of time will by chemical changes taking place in the plates, generate a current of electricity in a circuit, the current being in an opposite direction. The reader may form a fair idea of the principle, from a description of The Electrical Accumulator Company's storage cell. (See Fig. 81).

This cell is made up of fifteen plates, eight negatives and seven positives, and is especially adapted to isolated and central station lighting. The electro-motive force of the cell is about 2 volts. The internal resistance is extremely low, say from .001 to .005 ohm, and the range of the current large. The capacity of the cell in perfect condition is somewhat underestimated at 300 ampere-hours; 30 amperes, a safe working current, will last for over ten hours, with not exceeding 10 per cent drop in electro-motive force, or a less current will be supplied by the cell for a proportionately greater number of hours. A greater rate — up to 300 amperes — could also be obtained, but so great a strain upon this size of cell would injure the plates.

The chief advantage to be obtained from a storage battery system is to do away with overhead wires. Of course a power station is necessary where the batteries may be charged from dynamos, after which they are placed upon the car (usually underneath the seats), and the car is equipped with motors. (See Fig. 82.) Enough cells are placed on the car to run it about twelve hours, after which they

Figure 81.

must be removed and newly charged cells put in their place. Unfortunately storage batteries are very heavy, and their weight added to the weight of a car and its motors require more electrical power to propel them, besides much of the efficiency of the cells are lost by their heating from ohmic resistance, also from over charging, forming of foreign substances such as basic sulphates, etc., in the cells. Another serious fault is the buckling, particularly the positive plates which finally results in a short circuit and destruction of the plates. A new cell called the alkaline zuicate cell has recently been brought forward and which is lighter than the heaviest lead plates and is meeting with some success. The positive electrodes consist of porous copper plates which are formed by the compression of finely divided electrolytic copper upon a nucleus of copper gauze. They are surrounded by parchment paper cells to prevent any cupric oxide which is slightly soluble in caustic alkalies, but does not dialyze readily, — from becoming mixed with the potassium zuicate. The negative electrodes are made of amalgamated tinned iron wire gauze. Its E. M. F. is about .8 of a volt per couple.

Although a great deal has been claimed for the storage battery system it is a noticeable fact that at the present time there is not a road in the United States operated by this system. Several electrical companies are experimenting, among which is the Ford & Washburn Co. of Cleveland. An illustration of their storage battery car is given in Fig. 83.

The following account of their trial trip was taken from *Bubier's Popular Electrician* of April,'92: "Last month the Ford & Washburn Storage Battery Co., of Cleveland, Ohio, ran experimental trips with their storage battery car. Short trips were at first made along Woodland Avenue, between Willson and East Madison Avenues, and were quite successful. On Saturday afternoon at 8 o'clock the new car, loaded down with city officials, newspaper men and other prominent gentlemen, made a trip to Lake View cemetery and back. The start was made from the corner of Water and St. Clair streets, and the run to Lake View was made in twenty-six minutes. This is con-

FIGURE 82.

108   ELECTRIC RAILWAY ENGINEERING.

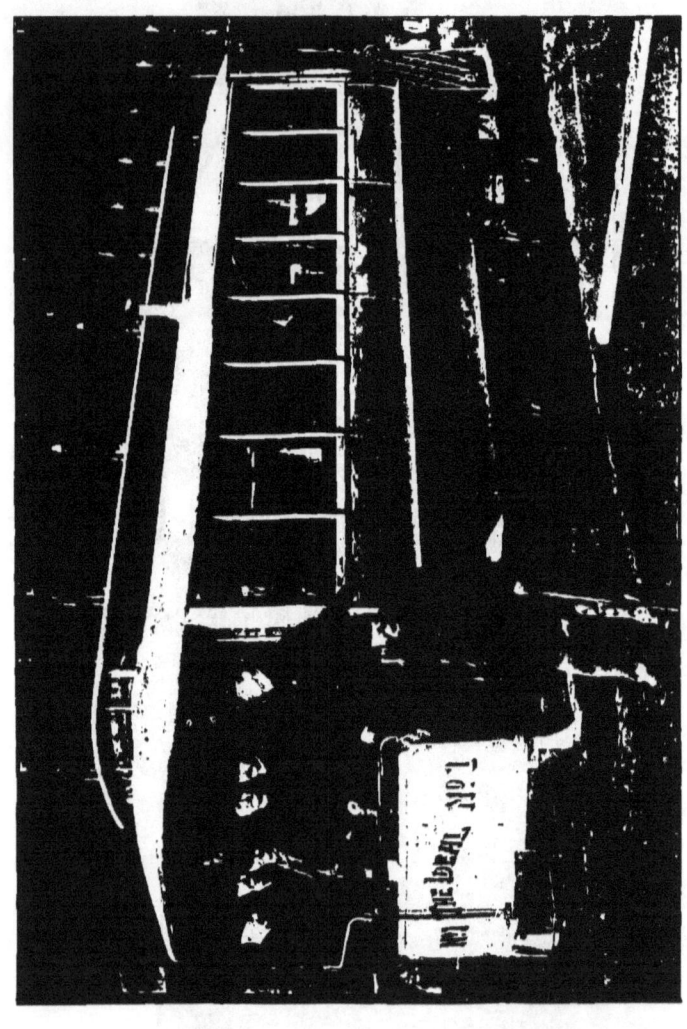

FIGURE 83.

sidered very good time, especially so as the car was delayed a few minutes at the Euclid avenue crossing of the Cleveland & Pittsburg Railroad. The ride was a pleasant one, the new car skimming over the excellent track of the East Cleveland road very smoothly. Curves were rounded with scarcely any jarring, and the grades between Willson avenue and the cemetery were ascended easily and without any apparent modification of speed. All the gentlemen in the party were unanimous in the opinion that the system is a good one and success is predicted for it. The motor is underneath the floor of the car, running lengthwise from end to end, and the batteries are placed on either side under the seats. The car can be made to go at almost any rate of speed, fifteen to eighteen miles an hour being the customary rate. We wish them every success."

In conclusion, it may be said, although much desired, the success of the storage battery system has not yet been established, all past and present efforts being only experimental.

CHAPTER XII.

SOME ILLUSTRATIVE ROADS.

EXPERIENCE proves that electricity is the best power for street car propulsion. Its advantages over the horse railway system are: its economy, its greater speed, and larger cars, with more seating capacity. Also, one motor car can be made to draw one or more cars (see Fig. 84), as the needs of the public require. Its use is therefore an advantage both to the company and the public whom they serve. To give the reader some idea of what has and is being done in this direction, a brief description will be given of some proposed roads and of some in actual operation.

Proposed System of the Paris Underground Electric Railway. — *The construction of the proposed underground electric railway for Paris will probably commence May 1, 1892, and it will require two years to complete it, according to the statement of the engineer, J. B. Berlier, of the Compagnie Les Tramways Tubulaires Souterrains de Paris.

As the question of rapid transit in New York city and Chicago is now exciting much comment, it will be of interest to many to see the methods to be used in the construction at Paris.

The total cost of the tunnel and equipment will be 54,000,000 francs, or 4,500 francs per metre for the tunnel, and an expenditure of 200,000 francs for each station.

The total length of the line first to be built is 6.1 miles (11 kilometres), and will extend from Bois de Boulogne across the city to the Porte de Vincennes. The entire system is divided into three lines; the first connects the Place de la Concorde to the Bois de Boulogne,

---
*Electrical Review.

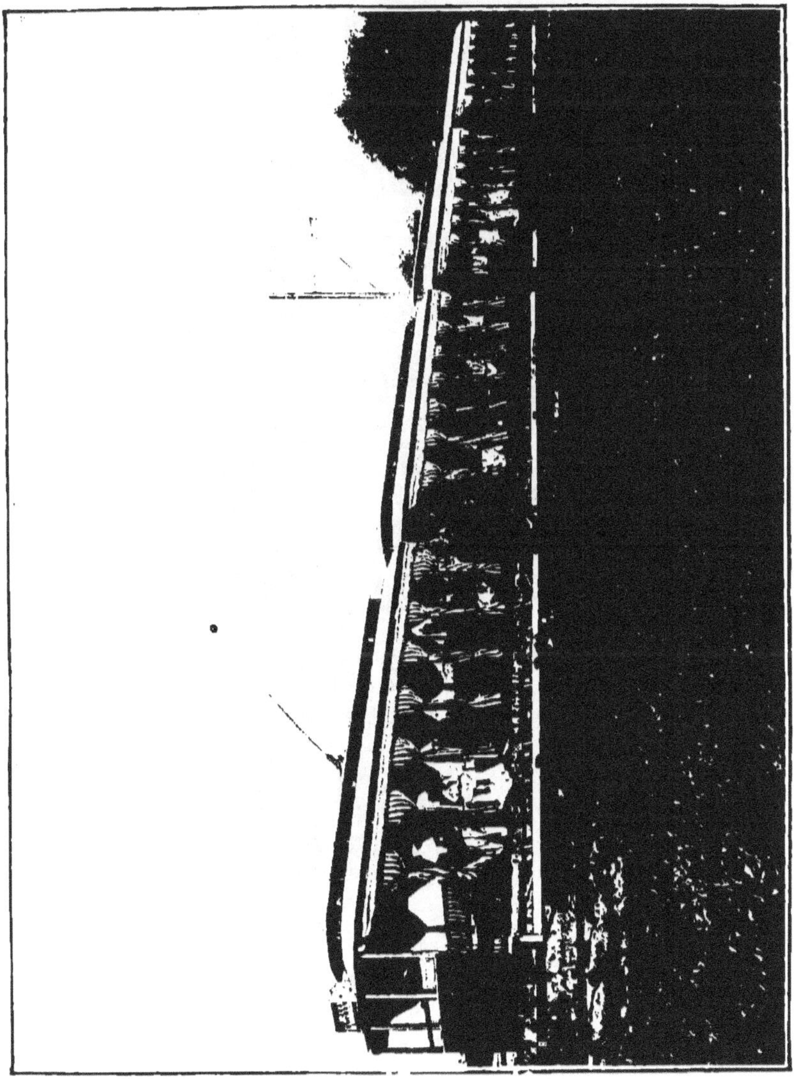

SOME ILLUSTRATIVE ROADS.

FIGURE 84.

FIGURE 85.

and is to have five stations; the second will start at Place de la Concorde and end at the Place de la Bastile, passing under Rue Royale and under the grand boulevards (Fig 85), with seven stations along the line; third line will extend from Place de la Concorde to the Porte de Vincennes, and will be connected to the first two. The whole system will be underground, except at the bassin de l'Arsenal which will be a viaduct.

The tunnel is to be water-tight and constructed of cast iron plates (1.5 metres) 1½ metres long, 50 centimetres (½ metre) wide and 2½ centimetres thick. These plates will be so put together as to

— Mode d'aération.

FIGURE 86.

form a tube 5.6 metres in diameter, and the tunnel will be constructed in the same manner as that at London, by hydraulic presses. The ventilation is accomplished by shafts placed along the tunnel at intervals (Fig. 86) of 50 metres, and Mr. Berlier claims that the movement of the trains will be sufficient to insure perfect ventilation. He also claims that many people will prefer traveling in the tunnel to being rattled about in the omnibuses used at the present time, for it will be very comfortable traveling, as the tunnels will be lighted throughout with incandescent lamps, the stations with arc lamps and the carriages with electric lights varying from 10 to 35 candlepower each, and the temperature in the tunnel will be more uniform than in the street, being cooler in summer and warmer in winter, while the only thing to contaminate the atmosphere of the tunnel will be the respiration of the passengers, as there is no smoke and no use of oxygen for gas or other lights. The stations are 25 to 30

FIGURE 87.

metres long and 15 metres wide (Fig. 87), and large staircases will lead to the street, while at the stations at Arc de Triomphe and Gare de Lyon there will be electric elevators.

The tunnel is to be placed 1.5 metres below the surface of the street as a minimum, and the maximum depth is at the Arc de

SOME ILLUSTRATIVE ROADS.

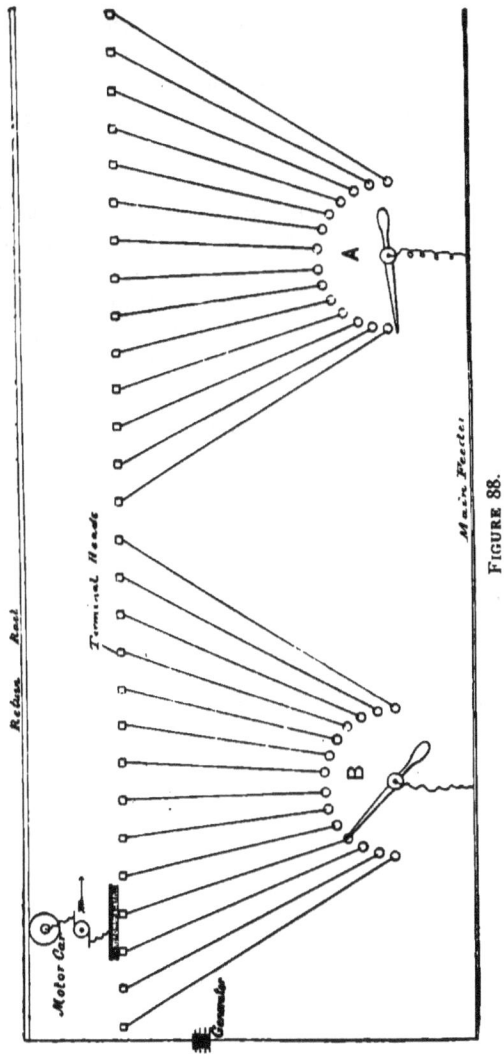

FIGURE 88.

Triomphe, where it is 18 metres below, and hence an elevator is necessary at this point. The heaviest grade is at the Place de la Bastile, where the road passes over the canal; at this point for over 20 metres the grade is six per cent or six centimetres per metre.

The central power station will be placed underground at the Place de la Bastile, and will have a total capacity of 4,000 horse power.

In a recent issue of the Electrical Age, N. Y., appeared the following description of a new system of electric street car propulsion, under the auspices of the American Engineering Company. The system is the invention of Mr. Granville T. Woods, who made an arrangement with the American Engineering Company to put his invention into practical operation. The relations between the parties have, through a misunderstanding, been severed, and Mr. Woods now claims that the company's possession of the patents on his invention were obtained in an irregular manner.

The system is known as the "Multiple Distributing Station System," and possesses features that are novel and distinct from anything ever devised for street car work. Between the rails, 12 feet apart, are laid iron blocks or "heads" about the size of an ordinary granite paving-block. Each of these iron blocks is connected electrically by an underground wire to the distributing station near by. The distributing stations are built in the form of lamp-posts. In each lamp post is placed the distributing apparatus, which consists of an automatically operated switch, which connects the feeder-wire with the "heads" as the car passes over each of the latter, and after the car has passed, the "head" becomes disconnected with the feeder and becomes "dead." In this way, only the "head" immediately under the car is alive as the car passes over it. The connection between the "head" and the motor is formed by contact-brushes attached to the car, and when the front brush strikes a "head" the rear brush is just leaving one.

In Figure 88, A and B represent two of these distributing stations, showing the contact-pointers, which are turned automatically. In the diagram at the left hand is shown the position of the motor car

SOME ILLUSTRATIVE ROADS.

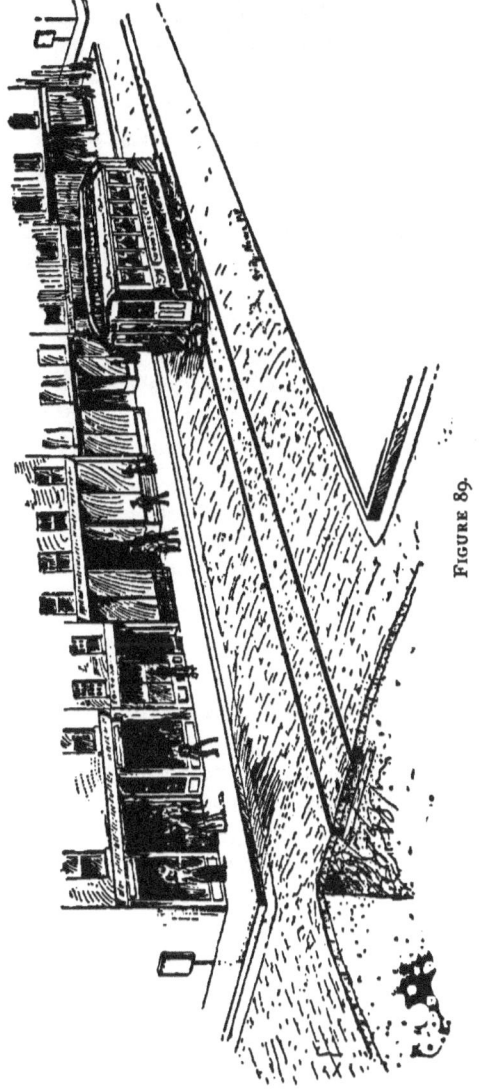

FIGURE 89.

on the "heads," and the "head" directly under the car, it will be noticed, is connected with the feeder through the automatic switch.

The main features of this system are that there are no wires exposed; no portion of the line is alive except the "head" beneath the car, and a block or derangement on any portion of the line does not affect the operation of the rest of the line. The system employing distributing stations located a block apart, provides a means for lighting the street along the line with electric lights, without having exposed wires at any point. The same motors and generators now employed on electric lines can be used on this system, the only expense being the installation of the underground wires and apparatus.

The test referred to, from all accounts, was very successful and made a very favorable impression. It may be mentioned incidentally that Mr. Woods is a colored man of extraordinary ability and intelligence. He was born in Australia, and came to this country when very young. He is the inventor of other electrical apparatus besides the one referred to above, and has received special training in electrical and mechanical engineering. Fig. 89 shows a general view of the system.

Electric Railway at Neversink Mountain, Reading, Pa. — The successful operation of the electrical railway up the famous Neversink Mountain, near the city of Reading, is another evidence of the rapid progress which electricity is making in the railway world. Considering the physical conditions, the steep grades and the weight of the cars on this line, it may be considered one of the most remarkable electric railway plants in operation.

The railway line, starting from the heart of the city, extends to the top of the mountain, which it overlooks, the total length of the road being twelve miles, including the road down the farther side. The ascent from Reading is made by a series of curves and one switch back. The grades are as high as 6.4 per cent, with scarcely a tangent on the whole line, from the time it leaves the city limits.

The road is used entirely as a means of reaching a number of

FIGURE 90.

different pleasure resorts, and to take advantage of the beautiful scenery which is afforded of the valley and the Schuylkill River, which winds around the foot of the mountain on the western side.

The road commenced operation during the summer of 1890. It is equipped throughout (with the exception of a short space within the city limits) with fifty-six pound "T" rails. The cars were equipped with Edison No. 6 double reduction 15 h. p. street car motors, each car employing two motors. Owing to the fact that the weight of the cars when empty is thirteen tons, and that they are often called upon to carry a load of 100 passengers up the very steep grades, it was deemed advisable in installing additional apparatus during the present season, to equip each of the cars with two of the new 25 h. p. single reduction motors, which are giving excellent results. The entire installation has been made by the Edison General Electric Co., who are to be congratulated on the success attained.

At present there are in operation six thirty-six foot cars, each weighing about thirteen tons. The cars used are of the Brill double truck pattern. The speed attained by a loaded car while ascending the 6.4 per cent grade is about eight miles an hour and twelve miles an hour on a 4 per cent grade.

One of the interesting features of this installation is the absolute lack of noise from the motors now in operation. Considering that the cars are exceedingly heavy and the grades uncommon, this feature of the road has won much praise. Part of this is undoubtedly due to the special oil boxes used with the gears.

The power station is situated on the Schuylkill River at the extreme end of the line. It contains two Edison eighty kilo-watt generators, driven from counter-shafting operated by two turbines. The weight of the cars, the type of the rails and the character of the roadbed closely resemble those of a steam railway line, and indicate that the Edison General Electric Co. do not intend to limit their operations to ordinary street car work.

The accompanying illustration of this railway conveys some idea

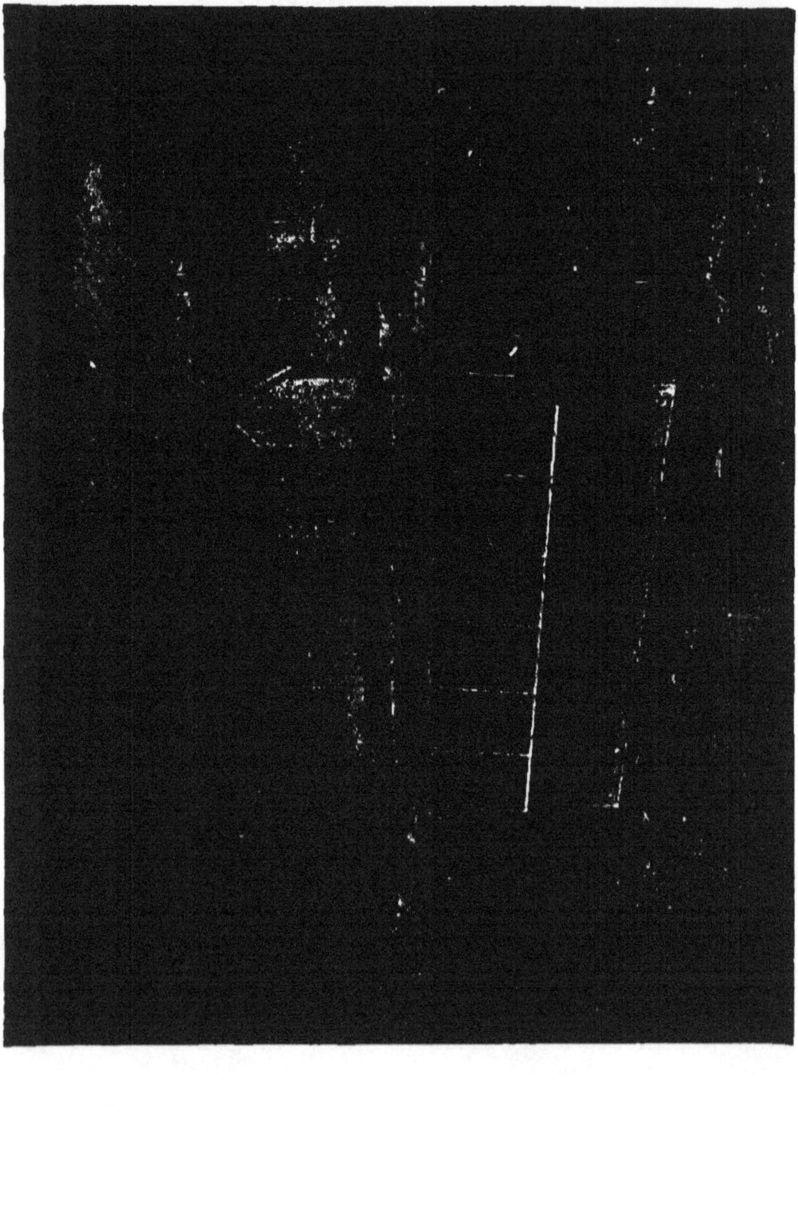

122 ELECTRIC RAILWAY ENGINEERING.

FIGURE 92.

of the difficulties to be overcome on account of the grades and curves. (See Fig. 90.)

We give an illustration (see Fig. 91) of one of the six-wheel electric motor cars such as are now used upon the West End Railway of Boston. The view is taken on the Tremont Street and Shawmut Avenue line. It has a seating capacity of sixty persons. This line uses the Thomson-Houston system, and each car is equipped with two of their new railway motors.

On page 122 (Fig. 92), will be found an illustration of Lynn & Boston Street Railway Thomson-Houston electric system in Lynn. The illustration shows a ·car ascending the grade on Rockaway Street, which is a very steep one. Cars are run up this grade at the rate of twelve miles an hour. Within the next year nearly all of the street railways in Lynn will be operated with electricity.

CHAPTER XIII.

SOME GENERAL REMARKS FOR MOTOR MEN.

THE substitution of electricity for horses on car lines has not lessened the amount of abuse to the motive power. The motors are unfavorably situated, being underneath the car in the dust, dirt, mud and snow. It is not many years since a noted electrician spoke of the commutator of a dynamo or motor as a most delicate piece of apparatus, that must be carefully shielded from all semblance of rough usage. If electrical engineers had not triumphed over these difficulties there would not be any electric cars to-day. Pails of water may be poured over a running motor without inflicting any discomfiture. Still, like all other machinery, electric motors have their ills and need appropriate remedies. When a slight accident has occurred to the mechanism of a car or motor, motor men generally pay no attention to it, and persist in keeping the car moving, at all hazards. Then if the motors absolutely refuse to move, the car is made to wait for another to push it along, or is drawn to the power house by horses. A better way of treating the same accident would be for the motorneer or conductor understanding the mechanism to locate the fault and make the necessary correction. Usually the remedies are remarkable for their simplicity.

The greater part of electric cars are controlled by means of a rheostat, and the explanations and suggestions given will first be for equipments of this kind. Trouble for cars may be located at any point from the place of contact of the trolley wheel with the overhead wire to the rails, or even further, when the car gets off the track, and we will consider the whole subject in an orderly manner suggested by the path of the current.

If the trolley wheel sparks and flashes, it may be due, in winter time, to the layer of ice that will gather on the lower edge of the wire. In summer, dirt and dust often gives imperfect contact. No remedies are advised for such causes. Oftener sparking is caused by the wheel slipping and jumping on the wire, instead of running smoothly. This is due to lack of oil on the wheel stud. Graphite bushings are usually inserted in the wheel hub to give self lubrication and offer a fairly good path for the current. This device would be perfect, if the dust and grit did not interfere. Some oil is necessary, but it must be used sparingly, as oil is an insulator, and the path for the current must be kept from all unnecessary impediments. Sometimes, but not frequently, an "open circuit" results in the wire that leads from the trolley base becoming detached from its clamps. Where the same wire enters the two main switches, a loose connection may occur, which may make a break in the circuit. More often the wires loosen under the car floor, where the location is not so easily detected.

In the fuse box or "cut-out" are two thumb screws for holding the fuse. When no current can be made to pass through the motors, after completing the circuit in the usual manner by the rheostat, always inspect the fuse box. The fuse may be blown, or the screws rattled loose.

The lightning arrester may sometimes get "grounded." After disrupting several lightning discharges, the parts are sometimes fused together, making a ground. In this case the ground wire may be cut or removed from its binding post. If it seems likely that there is an open circuit in the lightning arrester, wind some wire around the main binding posts so as to connect them together.

More trouble usually manifests itself in the rheostat than elsewhere. It is the most convenient place, and repairs to this part of the equipment are not expensive. Quite often the rheostat plays the trick of grounding the resistance plates on the iron containing-frame. In such a case the resistance in circuit is lessened, sometimes, there being no resistance at all in circuit, and the car will take a tremendous jump on starting. At other times the center stud on which the contact arm turns gets grounded with the frame, and it is impossible to shut the current off by turning the controlling crank. The motorneer should then open and close the circuit for starting and stopping the car by means of the main switch located over his head. The rheostat can still be used to govern the speed of the car if no other ground exists between the resistance plates and the frame, any such contact will readily show itself as a short circuit, accompanied by flashing and arcing of the current. In such a case the rheostat contact shoe on the movable arm should be moved entirely over onto its last contact (excepting the "loop" contact. which is the furthest position). The car can be controlled then by means of the main switch only. Violent shocks will be imparted by the motors as they start from rest, under the influence of the full force of the current. The shocks will be severe to the winding on the armatures, and to the teeth of the gears; this treatment should be permitted long enough only to get the car to the repair station. If the motorneer judges that the teeth of the gears are worn beyond the safe limit, it would be more advisable for him to await help from some other car. In case it appears that there is an open circuit in the car wiring, let the driver inspect the three wires that connect with the rheostat; a wire may have jarred from its fastenings.

The path of the current beyond the rheostat is divided, half the current going to each motor. Under ordinary circumstances the two motors are in multiple, that is, each one gets its current independent of any connection with the other. This is an essential feature, as it allows the car to be driven by one motor alone, if the other is disabled. If the motors were in series, and one should be cut out, the other would receive twice its proper potential.

The two field spools on each motor are usually in multiple; this allows finer wire to be used, and consequently more compact winding than if the entire current passed through each spool. A more important reason is that if a spool "burns out," it can simply be disconnected from the circuit, and the motor will run, though on account of its weaker field, more current should not be used as it would be absorbed and the remaining spool would be unduly heated. The current should be shut off from the motors when the car is running through puddles, as the continual splashing of water over the dampened connection boards often short-circuits a spool.

There are three cables connecting with each spool, two each for the "full field" when all the winding is in the circuit, the third leading to the very last contact plate of the rheostat and cutting out that part of the field winding called the "loop." The original function of the loop was to reduce the field magnetism to allow the motors to attain high speeds when there was almost no load. One of the conditions of non-sparking at the brushes is an intense field magnetism. When the loop is cut out the magnetism decreases, the current increases, and the chances for non-sparking are very unfavorable. Motorneers usually run their cars up grades and hills on the "loop" contact, but while the speed is slightly increased, there is about three times as much current absorbed as if the full field were in circuit. Unless a car is behind time, the loop should be used on level tracks only.

After the field spools, come the reversing switches. Street car motors are always series wound; hence reversing the current before it reaches the motors, would cause no reversal of direction of rotation. The switches must be placed between the field coils and the armature. The armature circuit is reversed, while the field remains always magnetized in the same direction. The armature could be made the permanent member, and the fields reversed, but on account of the loop contacts, the connections would not be so simple.

There are times, as in danger of a collision, etc, when a car needs to be stopped as quickly as possible. If the shutting off of the current and the application of the brakes will not do this soon enough, the reversing switch may be thrown over and the rheostat circuit as slowly completed as in starting a car. This means of stopping the car should only be in extreme cases, as the shock to both winding and gears will be severe, and even dangerous. Do not throw the reversing switch when the current is on. The arcing at the switch contacts would be liable to work damage, the excessive current might burn out the motors; at least the fuse would be blown, leaving the current wholly shut off from the car when most needed.

The armatures of the motors are next in circuit. A variety of troubles may occur in them. If a short circuit takes place, the smoke and fire will leave no doubt in one's mind that the motor is useless. It should be disconnected from circuit. This can easily be done by removing the brushes. Sometimes a ground is made and the driver cannot tell in which motor the trouble is situated. In this case, set the brakes and take out the brushes from one motor and let on the current gradually. If the fuse does not blow, the proper motor has been removed from circuit. If the fuse does blow, replace the brushes in the first motor, and remove them from the other, and the car can usually be driven by the first motor.

There is occasionally trouble at the commutator; the brushes spark badly. Before concluding that they are not set rightly, as regards the neutral point, clean off the dirt and carbon dust. Make an examination of the brushes; see if they are not too much worn, see if they are worn cornerwise, if so, put them in with the other end in contact. The copper plating will do less damage than the sparks. Often the carbon is not pressed against the commutator hard enough, the pressure springs may be broken or extended, this may be caused by the current taking the temper out of them. A block of wood under the presser will usually remedy this trouble temporarily by holding the carbon down and removing the springs from the circuit. An uneven commutator is the source of a great deal of sparking. Turning down a lathe until smooth is the proper remedy for this contingency.

Loose or broken connections between the armature coils and the commutator segments are of frequent occurrence. When the brushes touch the disconnected segment, a serious flash follows the break of the circuit. If the wire leading to the segment has been broken off short, it may have broken the continuity of the armature circuit ; in this case the armature will either turn weakly or not turn at all. When a broken connection of either coil is discovered, the motor should be removed from the circuit by taking out the brushes The commutator should be kept clean. Carbon dust makes a good path for the current, and any tendency to arcing around from one brush to the other may generally be traced to dirt. The neutral point on which the brushes should be set to avoid sparking, is, at best, found within narrow limits. The location of the yoke that supports the brush holders should not be changed from the marked position, without due care. The motors should be run at highest speed, the car stopped and the position of the yoke marked, then reverse the current, and run at full speed in the opposite direction ; if there is sparking in one case more than in the other, the position of the yoke may be changed, perhaps $\frac{1}{18}$ or $\frac{1}{32}$ of an inch. The direction and amount of movement should be determined by running the car back and forth, if possible, up and down hill. The shifting of the brushes in this manner should be done as a last resort, as often the sparking is due to the other causes that have been mentioned.

Sometimes the copper plating on the carbon brushes peels off and a corner or edge scrapes on the commutator, causing bad scoring and sparking. In this case strip the refractory piece of plating completely off.

The current passes from the armature back to the reversing switch, then to the iron frame of the motor which constitutes a "ground." If this ground connection loosens, an open circuit will result. The reversing switch should be examined occasionally to see if the contacts are good. Sometimes the jarring of the car rattles the switch blades out of contact, or so that there is "arcing" which quickly destroys the parts. The current may seem to come by fits and starts, causing severe jerking of the car. The icy condition of the track in winter will very often cause this trouble. In dry summer weather the dust on the rails may cause the same trouble.

Sometimes a car may be completely insulated from the rails. In this case the driver should not try to run the car on the "loop" circuit unless the car is behind time.

Cars should not be run for any length of time at slow speed, as the rheostat heats seriously, and the efficiency of the motors is reduced. The same time can be made with greater economy by letting the car run on its own momentum until the speed is slow, then speeding up, and again running by momentum, repeating this process as long as necessary.

All sorts of apparatus has been devised to obviate the use of rheostats on street cars. Many such contrivances work well during experiments, but cannot withstand the abuse of continuous service. Such arrangements usually make various combinations of the field magnet windings of each motor separately, and also of the motors as a whole. The fields are sometimes wound in three sections each, and some of the time these are in series, at other times in multiple. In some cases the ordinary resistance of the spools in series is not sufficient, and one or two sections are wound with German silver wire. This is an unwise method, as the heating of the fields is thereby increased and the efficiency of the motor reduced. It is better to have a small independent resistance to be in circuit only occasionally, — just when the car is started.

The Thomson-Houston Company has just perfected a controller which allows six combinations, as follows:

1st. Both motors in series plus 5 ohms resistance.
2d. Both motors in series, resistance cutout.
3d. No. 1 combination repeated.
4th. One motor cutout of circuit.
5th. Both motors in multiple with full field.
6th. Both motors in multiple with portion of field shunted.

The constructions of such controlling devices is attended with many difficulties. The change of combinations must be made at times and in such a manner as to avoid sparking at the contacts. Where sparking occurs blow-out magnets are used and are to some extent successful. If excessive sparking occurs, the driver should examine the magnets to see if they are still in circuit when the contacts over them are made.

Greater care on the part of a motorneer is necessary with a "controller" than with a rheostat regulation. The marks on the dial to which the handle bar must be moved should be carefully noticed, or contacts will be imperfectly made and arcing and burning will result. Sometimes spring-catches are provided for holding the bar of the handle in the right location. As there are more wires running to the controller box than to a rheostat, the dangers of loose connections and breaks are increased. All the cables and bindings posts are usually marked 1, 2, 3, etc., or A. B. C. etc., so that they can be coupled in their proper order.

CHAPTER XIV.

SOME GENERAL REMARKS FOR STATION MEN.

A SUBSTANTIAL and well cared for power station is a necessity to the success of any electric railway system. In our observation, as a rule, most stations are well kept and usually present a scrupulously neat appearance. Cleanliness is of great importance; much trouble is caused by its neglect. To keep everything dry is another rule to observe, as dampness very often causes grounds. All of the insulation upon wires, etc., about the station should be carefully looked after every day. (This may prevent a great deal of trouble.) " Want of care often causes as much trouble as want of knowledge," and if station men will bear this in mind, and heed it, they may save themselves many inconveniences and accidents.

A dynamo should be firmly set upon a solid foundation. If the foundation is poor, the vibration caused by the rotation of the armature may damage a dynamo in many ways. The iron base of a dynamo should be properly insulated from the foundation to prevent a ground. A dynamo should always be located in a dry place. All contacts and binding posts about the electrical apparatus of a power station should be examined, and if necessary be tightened every day. Fuses should also be looked after. A brush should never be lifted from the commutator while the dynamo is running, as it will cause an arc and make a bad spot upon the commutator. Only gentle pressure of the brushes upon the commutator is required. The brush holder springs should allow a certain amount of flexibility in order to prevent sparking of the brushes at the commutator.

The commutator is one of the most sensitive parts of a dynamo. It should always be kept smooth. When a commutator gets rough from sparking of the brushes, it may be made smooth by the use of emery cloth. Fine sandpaper may be wrapped around a block of wood and pressed against the commutator, taking care to raise the brushes before the operation. It should never be done while the dynamo is at work. Great care should be taken to clean off any emery dust that may remain upon the commutator, brushes, or shaft, as it will cut the surface of them for a long time and might cause serious damage.

For spots and grooves on the commutator there is no remedy but turning in a lathe. Files should not be used; it is quite impossible to produce a true cylinder with them. Much sparking at the commutator is generally a sign of overloading, or a short circuit in the armature coils. Keep iron and steel tools away from the dynamo, and never file near it. Use brass or zinc oil cans for lubricating, do not spill oil or water upon a dynamo, and have shields to prevent adjacent machinery from spattering oil upon it. Have a pair of bellows to blow the dust from the commutator and armature coils of the machine.

Oil is an insulator, therefore very little of it should be used upon the commutator, a few drops rubbed on with the hand is sufficient. When it becomes necessary to replace an old commutator by a new one, it may be done in the following manner: Carefully remove the armature from the dynamo and place it upon two wooden horses. Attach tags to the wires leading from the armature to the commutator, and mark them with numbers to make sure of the proper place of each wire to the corresponding bars of the commutator, discontinue the wires from the commutator by unscrewing the set screw, or where they are soldered by unsoldering them by means of acid and a hot soldering iron, taking care not to short circuit any parts of the commutator with any of the molten solder. Take the old commutator off, which is quite a difficult piece of work, clean the shaft and connections and put the new commutator carefully into position and connect the wires in proper turn with the corresponding copper bars of the commutator by the set screws or by soldering with hard solder.

If a dynamo refuses to work or the potential drops, see if your connections are right. For the sake of personal safety it is a good plan to wear rubber boots and thick rubber gloves when at work around a running dynamo. Rubber covers are now made for the handles of all iron and steel tools which every electrician should use. Keep the current indicator free from dust. If dust or dirt collects in jewel bearings the needle will act sluggishly. See that the jewels in the current indicator are not cracked or that the screws that hold the bearings do not get loose, for should this happen the needle will drop down from its bearings and refuse to register.

See that binding posts and all connections are kept perfectly tight, insuring good contacts of the instruments upon the switch-board. When placing voltmeter in position to take reading, be sure that it is on a level. When not on a level the needle will act sluggishly. A good plan when needle does not work freely, is to give the voltmeter a slight blow on its side with the hand.

When setting up an ammeter in the circuit, it is highly important to have it perfectly level and secure it firmly to the switch-board so that it will not easily be moved out of position. It should be placed far enough away from the dynamos so that the needle will not be effected by the magnetism from them. An ammeter that is not level will not register properly. It is important that the contacts of the ammeter should be kept tight, as a loose contact will cause it to heat and endanger burning out of its coils.

Never pound screws into a switch-board, always start them by first boring holes. Pounding around a switch-board is liable to do much damage to the instruments upon it.

The jaws of a lightning arrester must be kept clean and their proper distance maintained from each other. They should be inspected every day.

CHAPTER XV.

CONCLUSION.

WE have shown in the preceding chapters that the trolley system is the only one in practical use at the present time. It is a system comprehending many parts, each of which are essential to its completeness.

The power house and its apparatus, the motors, the line construction, etc., have all been described.

The development of this system was not the result of a few disconnected experiments, but of much study and thought, extending through a series of years.

The Conduit system, the Double Trolley, the Storage and Battery system have all been tried, and with one or two exceptions have been abandoned. We do not know of a single conduit system in operation in the United States today, and of only two double trolley systems, (one in Cincinnati, Ohio, and one in Camden, Pa.) Several Storage Battery systems are being experimented with, but the chances of success are against them. They require the constant attention of electricians to keep them in running order and may be so far considered commercial failures.

One of the best examples of the successful operation of the single trolley system is found in the History of the Construction and Working of the West End Railway of Boston, Mass., as given by Louis P. Hager, and published by him in 1892. Their main power station is the largest in the world. The boiler house covers a space of 161 feet long, by 85 feet, 10 inches wide. The boilers are arranged in twelve batteries, six on each side of the boiler room. Each battery consists of two boilers, which are of the Babcock & Wilcox water-tube type. The twenty-four boilers are capable of supplying steam to the aggregate amount of 24,000 horse-power.

The mammoth Reynolds & Corliss engines were built by the Edward P. Allis Company of Milwaukee, Wis. The fly wheel is 28 feet in diameter, 10 feet, 7 inches face, and weighs 80 tons. It is double-crowned and carries two double ply belts, each 54 inches wide and 150 feet long, to drive the counter shafting. The dynamos which generate the current are 118 in number and have an aggregate capacity of 15,260 horse-power. The West End Railway Company also have power houses in Allston and Cambridge.

*"The cars of the West End Railway Company at the present time number 2,131, all told, of which, according to the fourth annual report of the company, 1,662 are horse-cars, and 469 electric. As the electric cars are capable of accommodating, at a safe estimate, one-third more passengers than the horse-cars, when the electrical system is fully established on the entire lines operated by the company, the number of vehicles will be reduced in that proportion, thus affording a great relief in the matter of blockades, and consequently more rapid transportation to all points in and out of the city.

Although but a trifle over 81 miles of the 260 operated by the West End Company are equipped with the electrical system, some idea may be formed of the aggregate benefits which will result from its completion, from the fact that suburban property has already in many places appreciated over 100 per cent in the districts reached by the electric lines, and the people are so gratified with the change from horse-cars as to be unstinted in their praise of the almost magical transformation it has wrought.

The electric cars are beautiful specimens of the car builder's art; commodious in seating capacity; comfortable, not to say elegant, in upholstery; finely decorated inside and outside, and they certainly present a most imposing appearance, traversing the streets with the mysterious force which the Thomson-Houston Electrical Company has so successfully supplied in the over-head system of electrical propulsion."

* History of the West End Railway, by Louis P. Hager.

As was said before, all this was not accomplished in an instant, nor was it accomplished without many difficulties and objections. The public were naturally afraid of electricity, nor is this fear wholly dispelled today. Speaking of the danger of electricity, Prof. Elihu Thomson says:

"The growth of electric railways has undoubtedly been very rapid of late, and as you have heard stated here, there are something like fifty roads in operation at the present time (March, 1889). The experience of these roads has dated back several years. Some of them, no doubt, were crudely arranged at first, but the whole matter is becoming rapidly systematized and taking very much better shape. In order to convey electricity any considerable distance it is necessary that we provide conductors to convey the current. We must also use a certain pressure on the current, or the electricity will fail to be carried. In arc lighting this pressure rises to as high a point sometimes as 3,000 volts, and yet I have known men to come in contact with such wires, getting the full strength, and not be killed. There are a few other cases where people have met with fatal accidents by putting their hands on wires with a current of from 2,000 to 3,000 volts. By common consent, however, the electrical fraternity have dropped down to a voltage as low as 500 volts. That is the voltage which is now used on electric railways. The object of dropping the voltage is to get two things; that is, to secure safety, and at the same time secure freedom from the tendency of the current to leave the wire, either on the car or anywhere else. The desire is, of course, to keep that pressure which will transmit the current over the line. We could operate the roads with 1,500 to 2,000 volts, but that is not feasible or advisable ; we would find more difficulties in the construction of our motors. We are forced to keep the current down in pressure. A great deal has been said about the volume of the current existing on these lines. I say that it has nothing to do with it whatever. The volume of the current is nothing; it is merely the pressure which is to be taken into account; and this whole question hinges on whether 500 volts is a dangerous pressure or not. Now, it is a fact, as I am told, that the Western Union Company use on some of their lines in New York city more than 400 volts; they use dynamos for working long lines. I have heard of various instances where the leakage in bad weather has been so strong that the instruments were overcharged and the operators could not even adjust their instruments with the pressure being as great as that. They use the dynamos to replace a certain number of battery cells. The number which they would replace of the Gove type, taken as an example, would be about 240 battery cells. It does not seem to them that that voltage has any particular danger in it.

"They have substituted dynamos having a current of large capacity in place of their batteries, and still they find no difficulties with it. The pressure is not high enough to do any harm to the person. It is true that it is a pressure which will give a shock. Nobody denies that. Almost any pressure will give a shock, but the question here is whether it is capable of giving a fatal shock. I do not think any evidence has been produced here which shows that it can produce a fatal shock, or has produced a fatal shock, which is the important point. There are fifty roads in operation. We are prepared to produce testimony in regard to persons who have come in contact with a current and have not been injured more than to get an electric shock. I have occasionally touched conductors of very much higher voltage than these. I at one time caught hold of a conductor having a voltage of 10,000 volts for a few moments. I got a very severe shock, but it did not kill me. On one occasion I caught hold of an alternating current of 1,000, which you have spoken of as an exceedingly dangerous current, and that did not kill me. I do not say that I would voluntarily take hold of one of these conductors and take that shock, any more than I would go and have a tooth extracted without any reason for it. But I do say that the escapes from serious injury from much higher voltages than 1,000 volts are frequent. The voltage which is now used on electric railways has been reduced to that which has been agreed upon as the practical pres-

CONCLUSION. 139

sure to use, involving safety and efficiency throughout the whole system.

"A good many points have been brought up here which it is hardly worth while to touch upon but I am impressed with their contradictory character in many cases. Sometimes we hear of the impossibility of touching the wires and of the impossibility of firemen cutting the wires, because they are dangerous. We hear the statement made that firemen will receive shocks. All they have got to do is simply to have nippers with a wooden handle. They can cut any wire without any trouble. The wires can be cut very nicely by an ordinary pair of pliers without any danger whatever. I can say that there is not the slightest difficulty in removing all the wires in a very short time, if you choose, without any danger to people standing by.

"The danger to persons is absolutely nil. The conductor comes up from the top of the car and is heavily insulated. That is necessary in order to have it stand moisture. It would give out at once if we did not have the circuit on the car thoroughly insulated, covered in as fully as possible, to prevent it from coming in contact with the car track itself. In fact, a car body is constructed of material which will not convey a current of 500 volts at all. As to the danger to watches, although I have not heard any complaints of any trouble of that kind, I should say that there might be some little effect on watches not made non-magnetical; but the great watch firms now-a-days are making watches which will get over that difficulty. In fact, it is being regarded as essential that a man must be equipped with a non-magnetic watch in these days of electrical growth. I have in my pocket a time-piece of that character, which I can put in a dynamo with the strongest possible magnetic field and it will not affect its running at all."

In answer to the question by the legislative committee, of how the 10,000 volt current was taken by him, he replied:

"It was taken from an alternating machine working the primary of an induction coil and the secondary of the induction coil was taken into my hands by accident, and I got the full strength. The probable reason I survived was that I jumped away quickly. I said, "Here, you jump out of this, or you are dead." I knew 10,000 volts would kill me. It was capable of leaping from one wire to another, and at a distance of three-eighths of an inch, and without first contacting them. I got hold of the wires, and took the current in my body. The volume I had at that time, I was satisfied, was sufficient to kill me. But there was an enormous pressure, 10,000 volts; it had the power to jump through the air, and I got the full effect turned both ways. It went through me both ways fifty times a second, this way and that way, and the only effect I felt was that my arms were numb when I got away."

The Professor was then asked what would have been the effect had the current been 50 or 100 amperes, and he said he had never heard of such a case. He believed that a man receiving such a charge "would be instantly vaporized—would disappear in smoke. But 10,000 volts would not do that. It would require a lightning discharge. Cases have been reported," the Professor continued, "where men have been struck with lightning where they have been found considerably burnt by a very heavy lightning discharge passing directly through them, an amount made up of so much horsepower, so that it practically disorganizes and vaporizes the whole structure. A tree may be struck by lightning, and every particle of moisture in it may be vaporized in an instant, and the tree be exploded in every direction, but I never knew of a case of that kind in a man. We are getting beyond any ordinary voltage produced by any apparatus at our disposal that could produce an effect like that."

In regard to the safety of the electrical system in thunder storms, Professor Thomson stated what nearly every one has observed, "that as the number of electric wires increase in a city there is less trouble from thunder storms. That is, there are so many points of escape for lightning discharges that very few places in cities are struck and injured during a thunder storm. In the country, in a suburban district, a wire of course, might be struck, but look at the chances of its going to the ground! It is only insulated by a little porcelain knob from the side wires, which are often connected by a most complete circuit with the ground, and we also provide lightning arresters, putting them along the line as often as they may be needed, for carrying off any charge which could jump more than one-sixteenth of an inch. Lightning will jump a mile in many discharges. I would put a line of lightning arresters along the road which would carry off any current that can jump one-sixteenth of an inch, to the ground. Is not that sufficient protection? The fact is, if the lightning ever strikes a line it will find a number of points of escape, and will not affect the car, or go through the car; it will jump at once at the lightning arrester."

"Does it not happen in some cases that lightning flies along the overhead conductor, and is grounded without passing through the lightning arrester?" was asked; to which the Professor replied: "I should say some would pass through the motors, and in other cases the trolley wires would act as a first-class lightning rod to any person in the car. It would follow the conductor, which is able to carry an enormous current, and would go to the ground without entering the car at all. In other words, a man on a locomotive, with metal around him, carrying a circuit from the top of the cab to one below, is not going to be struck with lightning at any time, and that is the condition in all these cases."

To the question, "In case of a very heavy lightning discharge, wouldn't it fuse the motor wires?" answer was made that it would not. "It would simply jump when it got near the ground. Lightning doesn't have time to fuse that motor wire. It jumps to the ground when it is near the ground. It passes through the circuit of least resistance—the least inductive resistance, not the least electrical resistance. In other words, I must explain that a little further. Lightning acts differently from the ordinary electrical current. The lightning is discharged with great suddenness. Now, if I try to make it pass through a coil of wire, it will jump across a space in the air, rather than go through that coil of wire, which would be 700 or 800 feet in length; it would rather jump this space in the air and go to the ground that way. Suppose it comes down from the overhead wire, and finds the coils of the motor altogether too long for it, it would take the next nearest iron post, the nearest metal post, which is only probably a quarter of an inch away at any point, and will go to the track that way. In other words, it will shunt the motor, and we put on devices for that very purpose, to prevent injury to the motor. We put on a little jumping space, so that the lightning can jump that space and not damage the motor, because it may tear through some part of the insulated motor, and we therefore make a point for it to reach the ground easily."

An illustration of the safety of the overhead wires may be observed in the evidence of Josiah Q. Bennett, President of the Cambridge Electric Light Company. We quote from him as follows:

"We have never had a claim made against our company for any damage from any cause, nor has there been any injury to any employee from electric wires of any serious character, nothing from which any one has not in a short time recovered. I would like to state in this connection that our employees are insured by the Employers' Liability Association, for almost the same rates used in the most favored cases; that is for sixty cents a hundred dollars of pay roll per annum. They get the full indemnity, $5,000 apiece for every man. That shows that the Employers' Liability Association, who have made a most careful examination, were satisfied that the danger of death from currents of electric wires was very small indeed.

"We are running, and have been for a year, currents of 400 volts potential, for operating motors in Cambridge, and I have not heard of any injury or of the slightest trouble from that current. We are also running 500 volt currents, and we have not received any complaint about those. We are also running 2,500 volt currents, and in their operation it has been shown that the tendency is to throw off a person who comes in contact with the wires; he does not retain it long enough to receive any serious injury, and the only way in which we have known of a man's being hurt is from his being thrown from a pole and from concussion on falling to the ground, not by the current itself." Thus it will be seen that the dangers from using electricity is comparatively small, providing due care is used in its operation.

It is hardly proper to compare the horse railroad with the electrical system of street railway, as in almost every respect, the latter is so much the superior of the former. This is true of every department of the system, including its management and operation.

In the latter respects, superior qualifications are required on account of the greater responsibility attached to the positions. When we consider the high rates of speed which it is possible to maintain, the greater loads that may be carried, the additional weight and running gear, it is plain that much attention and care are required. Some knowledge of electrical science is also essential as a qualification to the most efficient service.

Experience shows that annoying derangements of the machinery are liable to occur occasionally, demanding knowledge of the wrong condition, and ability to overcome it for the time being. This knowledge should be possessed by the motorneer, who should be able to see, at once, the obstacle, and in many cases be able to apply a temporary remedy. This same knowledge should be possessed in a greater or less degree by all who have anything to do with motors or with the apparatus of the power house. Want of this knowledge has been the cause why some men who have been long connected with the former system have had to give place to men of less experience in railroading, but with better understanding of the wants of the new order of things.

As has been intimated, it is not expected that every employee will be acquainted with the details of the road and its equipments, from the power house to the rolling stock, but it is important that each man should be familiar with the details of his department, so as to be able to remedy many of the defects as they arise. In order to be this he must be a student; he must read, think and learn how to apply his knowledge to the practical purposes of his calling. He must be progressive in knowledge, to keep up with the progress that is constantly being made in electrical science, or he will sooner or later find that when some emergency arises he is deficient, and will not be able to meet it.

The following are some of the many advantages of the electrical system of street railway:

1. It is more economical.
2. It affords better accommodations to the patrons of the road.
3. It favors more rapid transit.
4. Cars can run up steep grades where it would be impossible to use horses.
5. It opens up suburban residences and enhances the value of property in these localities.

With all of the advantages of the present system, perfection is not claimed for it. While it has reached a state of commercial success, we can hardly conceive what may be some of its possibilities. Already one or two of the leading companies have in process of construction fast speed locomotives, which are designed to travel from forty to one hundred miles per hour. Should these experiments be successful, as they now promise to be, the steam railroad may, at no very distant day, be a thing of the past, which would mean no smoke, no cinders and faster travel.

APPENDIX A.

CHRONOLOGICAL HISTORY OF THE ELECTRIC RAILWAY.

THOMAS DAVENPORT, a poor blacksmith of Brandon, Vt., constructed what might be termed the first electric railway. The invention was crude and of little practical value, but the idea was there. In 1835 he exhibited in Springfield, Mass., a small model electric engine running upon a circular track, the circuit being furnished by primary batteries carried in the car.

Three years later, Robert Davidson, of Aberdeen, Scotland, began his experiments in this direction; his aim was to supplant the steam railway locomotive by the electric locomotive. With this view in mind he constructed quite a powerful electric motor, which was mounted upon a truck. Forty battery cells, carried on the car, furnished power to propel the motor. The battery elements were composed of amalgamated zinc and iron plates, the exciting liquid being dilute sulphuric acid. This locomotive was run successfully on several steam railroads in Scotland, the speed attained was four miles an hour, but this machine was afterwards destroyed by some malicious person or persons while it was being taken home to Aberdeen.

In 1849 Moses Farmer exhibited an electric engine which drew a small car containing two persons.

In 1851, Dr. Charles Grafton Page, of Salem, Mass., perfected an electric engine of considerable power. On April 29 of that year the engine was attached to a car and a trip was made from Washington to Bladensburg, over the Baltimore & Ohio Railroad track. The highest speed attained was nineteen miles an hour. The electric power was furnished by one hundred Grove cells carried on the engine. The consumption of zinc by the acid was very large and the jarring and shaking of the engine broke the jars in a shocking manner, which made the expense so great as to prohibit its application to commercial purposes.

The same year, Thomas Hall, of Boston, Mass., built a small electric locomotive called the Volta. The current was furnished by two Grove battery cells which were conducted to the rails, thence through the wheels of the locomotive to the motor. This was the first instance of the current being supplied to the motor on a locomotive from a stationary source. It was exhibited at the Charitable Mechanics' fair by him in 1860.

Another early inventor, Dr. Joseph R. Finney, of Pittsburg, designed and obtained a patent upon the transmission of electrical power from a dynamo by a wire stretched about twenty feet above and parallel with the tracks. This trolley supported a light wheeled vehicle which rolled along upon the trolley wire, from this a flexible metallic conductor made connection with the motor in the car, the return circuit being through the wheels and rails to the dynamo.

Mr. George Green, of Kalamazoo, Mich., was another inventor of the electric railway. He used a battery to obtain the electric power and it was transmitted through the rails of the track and wheels of his car to the motor.

In 1879, Messrs. Siemen & Halske, of Berlin, constructed and operated an electric railway at the Industrial Exposition. A third rail placed in the centre of the two outer rails, supplied the current which was taken up into the motor through a sliding contact under the locomotive. This current returned through the wheels of the locomotive, the outer rails serving as conductors to the dynamo. The car accommodated about twenty passengers, and the speed attained was about eight miles per hour. The circuit of three hundred meters was made in about two minutes.

In 1880 Thomas A. Edison constructed an experimental road near his laboratory in Menlo Park, N. J. The power from the locomotive was transferred to the car by belts running to and from the shafts of each. The current was taken from and returned through the rails.

HISTORY OF THE ELECTRIC RAILWAY. 147

Early in the year of 1881 the Lichterfelde, Germany, electric railway was put into operation. It is a third rail system and is still running at the present time. This may be said to be the first commercial electric railway constructed.

In 1883 the Daft Electric Co. equipped and operated quite successfully an electric system on the Saratoga & Mt. McGregor Railroad, at Saratoga, N. Y. The current was taken from a third rail in the centre, between the two outside rails. The locomotive used was named the Ampere. The line was about fifteen miles long over steep grades and the speed attained was about eight miles an hour. Mr. Leo Daft was the inventor of this system.

In the summer of 1884 Mr. Daft successfully equipped with electric apparatus a short road upon one of the piers at Coney Island.

On July 27, 1884, Bentley & Knight opened an electric railway system in Cleveland, Ohio.

The first overhead system in the United States was constructed in Kansas City, Mo. It was about half a mile in length and was constructed by J. C. Henry.

In 1885 Charles J. Van Depoele equipped a short road at Toronto, Canada. In 1886 he equipped another line at Windsor, Ont.

On April 27, 1887, the Daft Electric Co. opened an electric road at Los Angeles, Cal.

In 1880 Mr. F. J. Sprague equipped a short electric road in St. Joseph, Mo., and another at Richmond, Va. In this road thirteen miles were equipped and twenty cars were operated successfully.

This was followed by the Thomson-Houston Electric Railway at Crescent Beach, Mass.

October 31, 1888, the Council Bluffs & Omaha Railway and Bridge Co. was first operated by electricity, they using the Thomson-Houston system. The same year the Thomson-Houston Co. equipped the Highland Division of the Lynn & Boston Horse Railway at Lynn, Mass.

Horse railways now began to be equipped with electricity all over the world, and especially in the United States. In February, 1889, the Thomson-Houston Electric Co. had equipped the line from Bowdoin Square, Boston, to Harvard Square, Cambridge, of the West End Railway with electricity and operated twenty cars, since which time it has increased its electrical apparatus, until now it is the largest electric railway line in the world.

From the above brief historical sketch, it is seen that the electric street railway system has made advancement since its inception in 1835, until today, when it may be said to have passed its experimental stage, and become an assured success. Electric roads are now rapidly springing up in all parts of the world. That the electric railway has come to stay there can be no doubt, but of its future developments, who can tell? No doubt, electricity is in its infancy, and that, within a few years, still further marvels of this mysterious power will be witnessed.

APPENDIX B.

FENDERS.

ANOTHER important feature is a safety device known as a "fender." Its use is to prevent injury and loss of life of persons coming in collision with electric cars. This is accomplished by either throwing them aside or catching them in a net, where without its use they would be thrown under the car and crushed. So important is this car attachment that the Boston city government have found it necessary to appoint a special commission to investigate the relative merits of different fenders thus far in the field. The West End Railway Company, recognizing the importance of the need of this invention, have offered a large sum of money for the most efficient fender. The following is an account of experiments made in Boston, May 2d.

*There was a large gathering of experts and interested spectators at Grove Hall, yesterday, to witness another series of experiments with electric car fenders. Half a dozen different forms were exhibited, their respective merits being demonstrated by means of dummies representing a man, woman, and two boys.

The first candidate for favor was the "automatic," which has several devices to secure its operation independent of the motorman. It has a cut-off which will shut the current from the motor as soon as the fender is struck, a valve attachment to scatter grit on the rails, a swinging tripper of adjustable height, a fender dropping close to the rail at the touch of an obstacle, and also acting as an auxiliary brake. The fender itself is a net hanging straight down from the dasher to the rail, with an inclined scraper behind it to keep a body from going under the wheels. In operation the legs and arms of the dummy suffered severely, as the body was rolled over and over.

*Boston Herald.

The next was a wooden scoop with a rubber edge, set at an angle. In two trials it carried the body along with considerable bruising, but at the third trial, with a small boy, the legs got under the edge of the fender and were crushed.

The third was the Sullivan, the invention of a motorman on the line. It has been tried once before. It is a simple perpendicular form of wood, with a rubber cushion or edge, and is dropped by a lever by the motorman.

In each trial it pushed the body along before it, but like the two others it made no provision to spare the victim the concussion with the dasher which, in case of a swiftly moving car, would be fatal.

No. 4 was a wire net extending three feet in front of the car, like the Appleyard fender. It is quite elastic, and if a body falls into it he is safe, but if he is lying on the track the fender is liable to pass over him and drag him along very roughly.

The Campbell fender was next tried. It is also a scoop net of twine backed by spiral springs, making a cushion which does not allow the victim to strike the dasher. It runs close to the rails, so that it will scoop up a body lying down. It worked well at each trial. Mr. Campbell had a second device, a V-shaped scraper just in front of the trucks, the sides of which were fitted with wooden rollers, and its apex composed of spiral springs. In each test with bodies lying on the track, they were pushed along and thrown aside without going under the wheels.

No. 6 was a simple scraper or plough, with a straight "landside" over one rail, and a "mould board" extending backward diagonally to the other rail. This also rolled the dummy off the track.

The last experiment was with the Cleveland, which the commission so far approves as to recommend that 50 cars be equipped with it. It is an iron frame or shelf, projecting some 18 inches from the car and resting some 12 inches above the rails. With a standing figure it worked well, catching it and carrying it along, but it makes no claim to do anything for a body lying on the rails.

APPENDIX C.

METHODS OF ELECTRICALLY CONTROLLING STREET-CAR MOTORS.*

BY H. F. PARSHALL.

WHILE in many respects the controlling apparatus for street-car motors and the general requirements of the same do not differ greatly from some other cases, there are some features that demand the closest attention if the car is to be handled either efficiently or comfortably, so far as the passengers are concerned. While the number of methods proposed and tried in times past has been great, at the present time there seems to be sufficient agreement between the principal designers and sufficient data at hand to warrant the writing of a fairly comprehensive paper on the subject.

The problem of controlling the motors is probably the most difficult one in the whole range of street-car work, and in no small degree determines the electrical design of the motors; or to be more specific, to start a car under any given conditions of track a certain torque is required. Beyond a certain limit, fixed largely by the convenience of passengers, this torque cannot be exceeded. The smaller the current with which the motor is able to develop this torque, the smaller the rheostat or other starting devices may be, and the more efficient the car equipment. Should the motor, therefore, be incapable of developing a comparatively powerful torque per ampere, the amount of energy dissipated either in the magnetic windings, armature windings or rheostat becomes excessive, the results being the more or less rapid deterioration of these parts.

It may not be out of order just here to discuss the design of the motor with reference to getting this torque most efficiently. The average horse-power exerted by a street-car motor at the car wheel

---
* Paper read before the American Institute of Electrical Engineers, New York, April 19, 1892.

probably does not exceed 20 per cent. of the maximum it is expected to do in starting the car under the various conditions encountered. Now, to get the highest efficiency from a motor run under these conditions, it is necessary to get the highest possible efficiency at that horse-power at which the greatest amount of work is to be done, and inasmuch as the loss in the conductors for this average horse-power is necessarily low (otherwise the motors would burn out in doing the maximum work to which they are subjected), the question does not resolve itself into how to get the least possible motor resistance of armature and magnets, but, rather, how to minimize the constant loss of hysteresis, eddy currents and friction. While all of these losses vary somewhat with the speed in series-wound motors, the variation of these losses is not great, since for an increased speed there is in general a diminished intensity of magnetization and pressure. To render these losses a minimum, and at the same time to get the requisite torque to handle the car efficiently, there is but one solution, that is to put the maximum number of turns on the armature compatible with good running as to heating and sparking.

While the truth of these statements may be more or less apparent to all when stated in plain terms, but little attention was paid to this matter in the earlier motors designed. The numerous measurements made, however, have so uniformly been in favor of motors with comparatively a large number of conductors on the armatures, that the importance of this matter is now pretty generally understood.

This agreement as to the general design of motors has in no small way been influential in bringing electricians into agreement as to how the motor should be controlled, since with an armature of a comparatively large number of turns, less turns are required in the field magnets to produce a given torque with a given number of amperes. The function of the magnets, therefore, has become of less importance. It is always, however, to be borne in mind that, other things being equal, the motor with the greatest number of turns in the magnets will develop the greatest torque for small currents. With a given electromotive force acting on the armature circuit, and a given torque developed by the armature per ampere, it does not matter, so far as efficiency is concerned, whether the difference in electromotive force at the armature terminals and the line is due to drop in external resistance, or to drop in the magnets. This point determines, once for all, that motors with commutated fields are not necessarily more efficient than other motors.

The particular advantages of the commutated field method are, that with a limited number of pounds of copper, or in the case of street car motors, with the limited space available for field-magnet windings, it is possible to adjust the magnetizing force of the field coils, so that the rate of doing work of the motors may be made to correspond with the rate this work is required by the car for the various speeds and conditions of track. This adjustment may be made for any size of motor with any required degree of precision by varying the number of magnetic coils. To increase the range or precision it is only necessary to increase the number of coils. In practice it has been found that this number could not be very great, otherwise the car wiring becomes too complicated and too expensive. This same holds true of the controlling-switch. Three magnet coils or sets of magnet coils seem to be the practical limit, since there is a general agreement between street-railway managers that the present number of magnet connections (6) should not be increased, and even with this number there is occasional trouble with broken wires or terminals. With a 51 horse-power motor it is possible with three sets of coils to run under most conditions met with in practice without employing external resistance. It is occasionally necessary, however, when the car is to be run at two or three miles an hour to make use of the resistance coil that is ordinarily used only when starting. With 25 horse-power motors it is necessary, with three sets of magnet windings, to make use of this resistance coil very considerably in ordinary practice, since without this it is not possible to get a speed of less than one-third maximum speed of the car, which is generally taken to be about 18 miles an hour.

The range of speeds without the use of a rheostat is determined by the limit to which it is safe to heat the magnets. The temperature of the magnets should not in any case exceed 65° C. This would put the increase of temperature at about 30° C. This increase corresponds to an average loss in the magnets of about 0.3 of a watt per square inch of radiating surface. For the few seconds generally taken to start the car the loss may be as high as two watts per square inch without dangerous heating. Experience, however, has demonstrated that to exceed this limit, even for very short periods, there is considerable risk. Having the maximum number of watts that may be dissipated in the magnets, the series resistance of same may be calculated from the properties of the motor on the supposition that each ampere taken by the motor produces a certain number of pounds pull at the periphery of the car wheel. In a well-designed motor with commutated fields it is easy to get from 35 to 40 pounds pull at the periphery of a thirty-inch car wheel with the coils in the series position. These coils are either wound side by side, or one on top of the other, according to the necessities of the case as determined by the general design of the motor. Experience has shown the advisability of winding these coils in independent spools whenever the general design will permit, since in the case of trouble it is cheaper and easier to replace the damaged coil, and there is less liability of crosses between the coils. As an example of a design that has been found to give general satisfaction in practice, I give the following figures from a series of tests made on a Sprague No. 6, S. C. motor, the magnetic data of which has already been published by myself : *

EFFICIENCY ON STREET CAR MOTOR NO. 6.

Brake H. P.	Efficiency.	Speed.	Remarks.	Res.
14.3	87 Per Cent.	1110	3 coils in parallel.....	0.8
11.8	87 "	1174	3 " "	.8
11.4	86 "	1184	2 " " ....:	1.4
9	84 "	1309	2 " "	1.4
8.25	82 "	955	2 in parallel, 1 in series.	3.24
6.35	79 "	1040	2 " 1 "	3.24
4.9	72 "	1070	2 coils in series	4.86
3.9	70 "	1014	3 " "	7.42

* Transactions, vol. vii. p. 218.

It is to be noted especially that it is possible to get an approximately constant speed with a wide range of loads, and yet have the energy dissipated in the magnets remain approximately constant, and that it is possible to get a torque corresponding very approximately over a wide range, to that required to propel a car under conditions met with. This is the solution of the question how to get the highest efficiency. For instance, suppose a car is to be run between two points in a given space of time, and this is not an infrequent requirement, and that the magnet windings of the motor are such that either the car runs the distance in too short a time, or in too long a time, it will be necessary then to accelerate the car for a time beyond the limit required, then to allow it to slow down, then to accelerate it again, or go through some such cycle of operations to get the required results. More power will be required with such windings than when such a torque can be had at the motor, that will produce the required speed by an approximately uniform acceleration. To get the same results given above for the No. 6 motor, with the magnet coils arranged in loops instead of separate coils, would require upwards of three times as many pounds of copper as was used in the present case (110 lbs.). This motor was designed to give a maximum car speed under ordinary conditions of from 12 to 15 miles an hour. At present it is thought advisable to have a maximum car speed of from 18 to 20 miles an hour,† since numerous measurements have shown the economy of running street cars at as high a speed as the conditions of track, etc., will permit. In a series of measurements made by myself it was found that the watt hours per car mile decreased very considerably with the speed of the car up to 30 miles an hour. To get this high speed (20 miles per hour), it has been found necessary to vary the proportions of the magnet coils from that given in the above for the No. 6 motor. Thus for a single reduction 15 horse-power motor the resistance of the last coil to be turned from series to parallel is only 15 per cent of the total

---

† All car speeds are quoted for straight and level tracks. These, when calculated for a new motor, are determined from the speed and horse-power curve of the motor, assuming the resistance to be 30 lbs. per ton. The methods of measuring these speeds are in general such that the probable error is too great to determine the percentage slip of the wheels.

resistance of the magnets, and the turns of this coil only 20 per cent of the turns in the other two coils. The reason for putting this low resistance coil inside, is to get the greatest number of turns when the coils are all in series, and the least resistance when the coils are all in parallel. Further, under ordinary conditions, this coil has the least expenditure of energy in it, and the least radiating surface. With a winding of this proportion, it is necessary with 15 horse-power motors to use an external resistance of 6 or 8 ohms. With 25 horse-power motors an external resistance of from 10 to 12 ohms is required. This resistance should be so sub-divided that there is not more than 20 volts E. M. F. between adjacent contact pieces, and so proportioned that the increase of temperature is not in any case above 150° C.

A method that is receiving a great deal of attention now is that known as the "series parallel method." While it has not yet been introduced very largely in practice, numerous experiments have indicated the desirability of doing this as soon as some of the troublesome features of the switch have been overcome. The method of operating is as follows:

In starting, a rheostat of from 8 to 20 ohms is used, according to circumstances, in series with the motors, which are in series with each other. After this resistance is thrown out of circuit the magnet coils of one of the motors are short-circuited, a section at a time. To make the start smooth, three or four coils at least are required. The magnet coils being short-circuited, the armature is then short-circuited, and the magnet coils thrown in circuit simultaneously with the armatures being thrown in parallel. It is just at this point where the difficulty with the switch has been encountered, since either the switch has to be operated with great rapidity or the contacts act in perfect unison, otherwise unpleasant results as to short-circuiting occurs.

The advantages of the method are that a very wide range of speeds are obtainable at a comparatively high efficiency, and that the energy required to be dissipated by the rheostat is small for the low

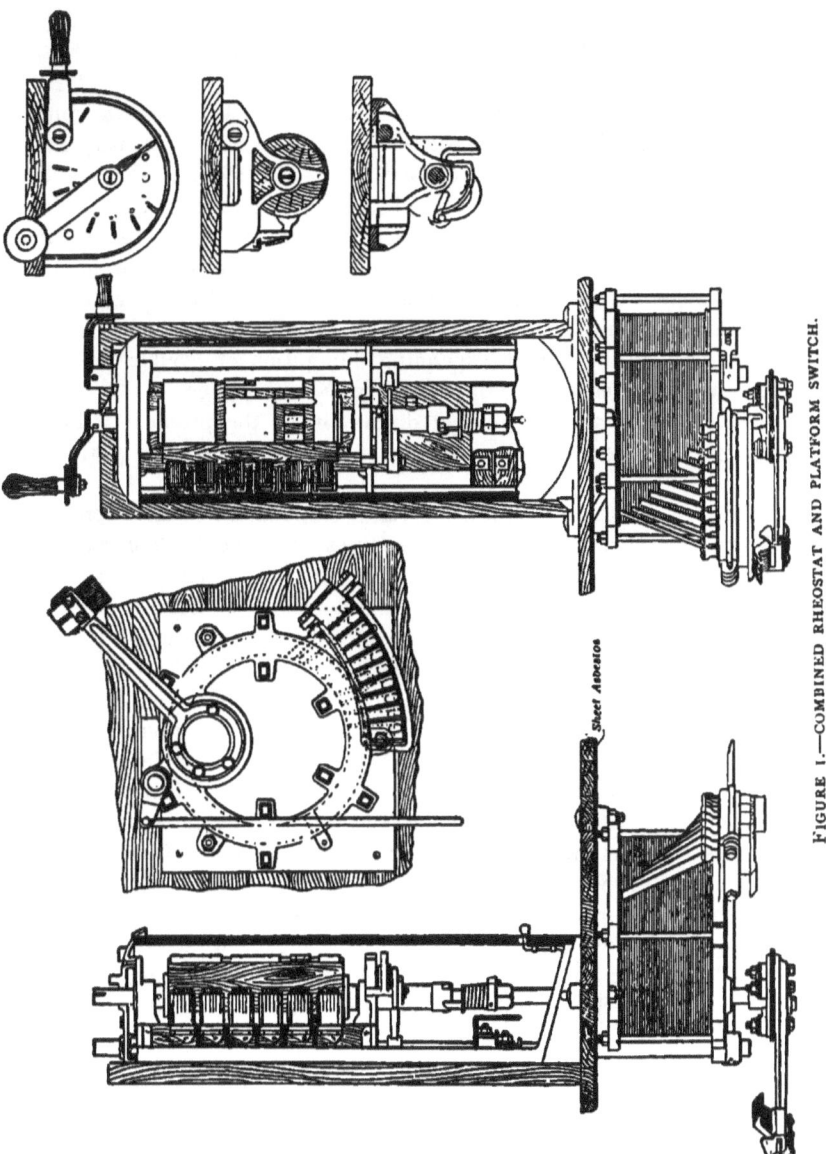

FIGURE 1.—COMBINED RHEOSTAT AND PLATFORM SWITCH.

speeds frequently required in city practice. This lessening the duty of the rheostat is a very important point, since as yet it has been found exceedingly difficult to construct a cheap rheostat that could be placed under the car in the small space available and dissipate so large an amount of energy as is required when the car is to be run for a considerable time at a speed as low as two or three miles an hour. Any method of control that has lessened the energy to be dissipated in the rheostat has in general been considered with favor, since there has been a corresponding diminution of trouble in each case that the energy to be dissipated has been lessened.

Having now given a general discussion of the problem a brief description of some of the apparatus recently devised may prove of interest.

Fig. 1 shows the general design and arrangement of an improved form of platform switch, which combines both the "field commutation" and the "series resistance" methods of starting cars. To start the car, the switch handle is turned from the position marked "off" with a counter-clockwise movement; this movement carries the arm of the rheostat, which is placed under the switch, around and over the contact segments, so that the resistance is gradually cut out of circuit. After the contact arm has been carried around to 135 degrees and all the resistance has been cut out, it is released from the cylinder shaft and left locked in this position. A further movement of the switch handle then affects only the cylinder, and commutates the sectional windings of the field magnets of the motor from series to parallel in the usual way. In stopping the car the field coils are turned from parallel to series, the resistance coil is then again put into circuit and the circuit broken when the contact lever leaves the last segment of the resistance coil, and not, as hitherto, upon the cylinder contacts. The only caution to be observed in stopping is to see that the switch handle shall be turned to the position marked "off," for the motors are reversed by means of a separate reversing switch placed under the car and operated by a lever connecting with a separate shaft in the controlling switch case. The shaft of the platform switch interlocks with this reversing shaft in such a manner that it is impossible to reverse the motors until the cylinder is in the "off" position. The use of this separate controlling switch has been objected to, but to combine both the advantages of the rheostat and commutated fields the switch mechanism becomes too complicated and the switch too large to have the reversing performed by a reverse movement of the controlling switch handle.

The cylinder plates and contacts are made of thick iron stampings, as experience has shown that iron is more durable than brass for this purpose. The burning, due to the formation of arcs, does not have so much effect upon iron as it does upon brass, and there is more certainty of good contact. The contacts on the cylinder consist of a number of stampings arranged in a brass frame, each stamping making an independent spring contact with the switch cylinder. The rheostat employed is built up in a circular form from a large number of flat rings stamped from thin iron sheets. The rings are built up in the form of a cylinder, each ring of iron being separated from the adjacent rings by a ring of mica, except at the point where it makes contact with the ring on the other side of it. Instead, however, of being arranged in a continuous spiral circuit, the coil is divided into a number of parts so arranged that the direction of rotation of the adjacent spirals is reversed, this being done to make the inductance of the coil as small as possible. A coil wound up in a continuous spiral having a mean diameter of 12" and a radial depth of 1", 6" long, and composed of 400 plates, was found to have an inductance of 40 mili-henrys. The coil was then wound up in 12 sections, the direction of each section being reversed, and the inductance in this case was found to be 8.5 millihenrys. These sections are stamped from different thicknesses of metal, so that those coils which are in circuit the shortest time and have the least current to carry are of highest resistance and least ampere capacity, and those that are liable to be in circuit for some time are thicker and have less resistance and greater ampere capacity. Copper connections are made at different points in the coil, all these connections being brought to a number of small iron contact pieces fitted in a spiral form and arranged so that the switch contact lever can slide over them. The contact pieces are insulated from the frame with sheet mica and from one another with small slate slabs. The rheostat is entirely fireproof and can expel with safety the heat evolved within it under all ordinary circumstances. As a point of practical importance it is, however, advisable to place a sheet of metal and a layer of asbestos paper between the rheostat frame and the car floor. This will prevent any danger from fire, either from heating or sparking, should such occur. It is to be noted that the general design of this rheostat is such that those parts having mechanical functions and energy-dissipating functions have been separated as much as possible. Of course the mechanical functions of a rheostat are more or less limited; it is evident, however, this effort is in the right direction. It is with respect to this particular point that the rheostat has a decided advantage over any form of mechanical clutch in starting a car. The clutch, of course, has its advantage in starting quickly bodies that have a great amount of inertia. In ordinary practice, however, the amount of energy dissipated in a clutch is approximately equal to that necessary to dissipate in a rheostat, but the clutch has in addition to its energy-dissipating function, a very exact mechanical function, and these two functions are interdependent on the same wearing parts. For this reason, if no other, clutches have not yet been made to compete favorably with rheostats.

Fig. 2 gives a diagram of the car connections for this switch. It will be seen that the current from the trolley wire first goes through the field coils and switch cylinder for commutation, then through the armature and reversing switch, and thence through the switch contact lever and resistance coil (in starting) to ground. It will be noticed that by use of the separate reversing switch the armature wires and field wires are each kept separate and distinct from one another. Formerly there was considerable trouble from the break-

ing of these wires, especially where the wire entered the brass terminals at the various terminal boards. This has been almost entirely obviated by using 49-strand cable wherever wire was subjected to bending.

In some cases the construction at the platform ends is such as to make it inconvenient to place the rheostat used with this form of switch immediately underneath the cylinder. This is the case when

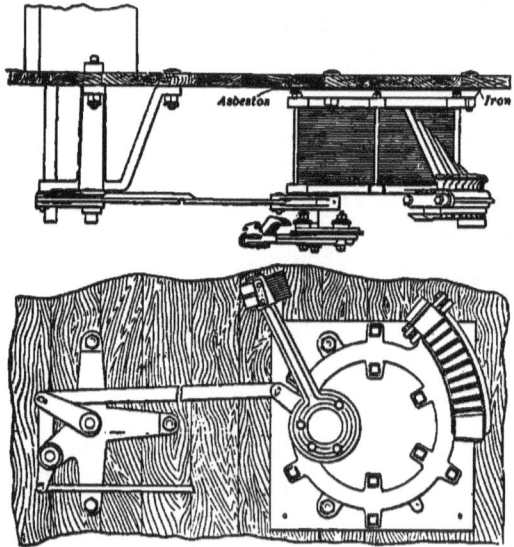

FIGURE 3.—DETACHABLE RHEOSTAT.

certain kinds of draw-bar or step constructions are used. In these cases a modification of the switch arrangements is made so that instead of the rheostat a light frame is placed directly under the cylinder. This frame serves to support the switch-shaft, upon which is placed a crank connecting with a bar, which is carried off to the rheostat contact lever. With this arrangement the rheostat can be placed under any convenient part of the car flooring and operated as well as when directly under the platform.

Figs. 4 and 5 show general plans of a car switch designed to be placed under the car and about half way between the motors, when the car construction permits. This design, while open to the criticism that the switch is somewhat inaccessible for inspection, meets the demand that has sometimes been made when it has been thought the space ordinarily occupied by the platform switch could not be sacrificed. The principle is the same as the platform switch already described, but it is modified in form and shape to suit the particular condition under which it is to work, and it is to be noted that the mechanical adjustments required are much more exact, otherwise there would be considerable burning of the contacts, since the motorman would be unable to tell whether or not the switch contacts were on proper positions. The rheostat is arranged in sections and connections brought from them directly to cylinder contacts. A cylinder is used to commutate both resistance and field magnet coils.

An important point that has been attended to in this switch is the breaking of the circuit on a separate switch instead of on the cylinder. A snap switch, of the knife blade pattern, is employed to break the circuit at four points. It operates in connection with the cylinder shaft, to which it is connected with a special locking and releasing gear of similar design to that shown in Fig. 1. The first movement of the cylinder shaft closes the snap switch and completes the circuit through the coil. Further movement then disengages the snap switch from the shaft (leaving it closed) and the different commutations are effected. When breaking the circuit the snap switch is again brought into action.

When this form of car-controlling switch is used, the platform lever is fitted at its lower end with a bevel gear wheel meshed into another gear wheel placed on the cylinder shaft. When necessary an extension shaft fitted with one or more universal joints makes connection between the platform lever and the cylinder shaft. When this switch is placed in the middle of the car, the amount of car

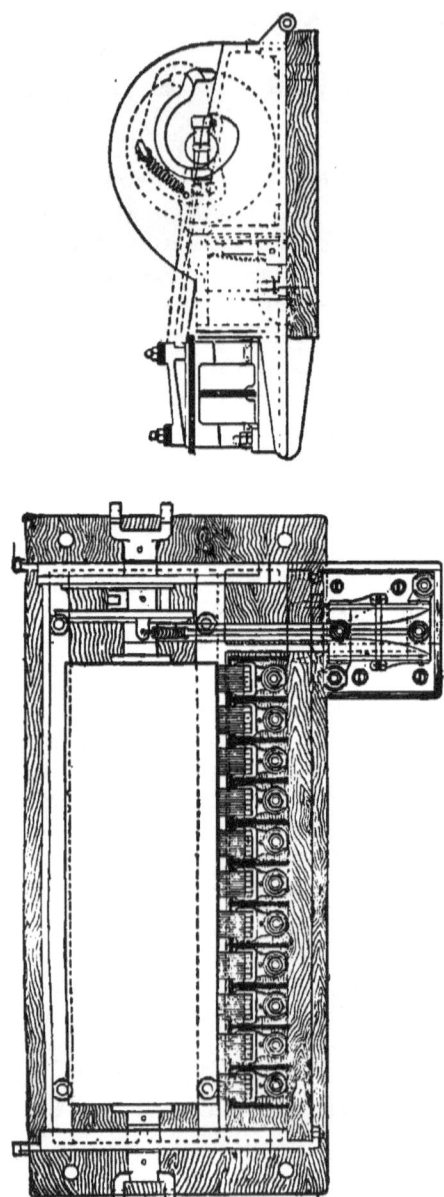

FIGURES 4 AND 5.—GENERAL PLANS OF SWITCH WHEN PLACED UNDER THE CAR.

wiring is materially lessened and the car inspection made more easy.

With reference to controlling switches in general, it is evident that a great number of designs may be prepared that will give approximately the same electrical results in point of efficiency. In deciding then upon the merits of a new design of switch, the commercial factor relating to repairs has therefore to be very largely considered, and had designers been able to guide their work more closely from the balance sheets of railroad companies, when such had been properly kept, instead of conforming to popular notions, very much more progress would have been made in this line during the last few years.

In closing this paper it might be well that I should remark that my experience has been largely confined to what is known as the commutated field method of control, and that I have naturally expressed many of the qualities of other methods in terms of this method. If these expressions are not judged satisfactory, I leave it for those who have had a similar experience with other systems to express in their criticisms the qualities of the commutated field system in their own terms.

NOTE. — The illustrations used in this article were kindly loaned to us by the Electrical Age Publishing Co., New York City.

APPENDIX D.

RAPID TRANSIT.

LESSONS FROM THE CENSUS.*

BY CARROLL D. WRIGHT, A. M.

WE have seen that the population of cities is rapidly gaining in proportion to the increase of population in the whole country, and also that this growth in cities is largely suburban in its character. The suburban growth is fed from without and from within. As business is extended, and the room and area formerly occupied by people are taken for great mercantile houses and for manufacturing, the population of such areas is sent out to the suburbs of necessity, while many seek suburban residences as a matter of choice. From without the suburban population is augmented by the rush to cities from the country. Owing to the improvement in methods of agriculture, by which production from the earth becomes in some sense a manufacture, a less number of persons is required for agricultural purposes than of old. The question is often asked why, if population increases, there is not an increasing necessity of supplying food products; and if there is such a necessity, why can great numbers be spared from the rural districts to engage in the business undertakings of the cities? Improved methods of production offer an answer to this question, the result being that the labor of the country not being in so great demand, even to supply the vast increase required in food products, seeks remunerative employment in centres of populations. As the contraction of labor through invention goes on, the expansion of labor through invention grows to a greater extent; and it is probably true that through inventions, or through great industries which have come into being in recent years, a larger number of people are employed relatively than are deprived

*Abstract from *Popular Science Monthly*, April, 1892.

of employment through improved methods. The great industries associated with electrics, railroad enterprises, the building of new kinds of machinery, and the absorbing in various ways of laborers in occupations not known until within a few years, enables manufacturing centres to furnish gainful work to those coming from the country, where, relatively speaking, they are not needed. These people take up their residence in the suburbs, though they may find their occupations in the crowded areas of the cities themselves. The question of rapid transit in cities, therefore, becomes one not only of great interest in the study of the movement of population at the present time, but one of prime necessity for the consideration of municipal governments. It is something more than a question of economics or of business convenience; it is a social and an ethical question as well.

The bulletins of the census furnish, to some extent, the statistics relating to rapid transit in cities, and of the relative economy of different motive powers used on street railways. These bulletins have been prepared by Mr. Charles H. Coolley, special agent for rapid-transit facilities in cities, under the immediate direction of that skillful statistican and economist, Mr. Henry C. Adams, special agent for transportation, and from them we learn the growth of rapid-transit facilities during the ten years from 1880 to 1889, inclusive, in cities having over fifty thousand inhabitants. The special experts have selected cities on a basis of an estimate of population made at the time the compilation of the tables was begun.

The full reports of the statistics of the equipment of all roads furnishing rapid-transit facilities, and of their operations for the single fiscal year ending 1890, are being collected, and the census authorities will present them in future exhibits.

Prof. Adams announces, and with truth, that street railways have never before been brought within the scope of the census statistics of transportation, and he points out the peculiar difficulties which were met with in collecting the facts already presented. Some of these difficulties arose from the ambiguity of designation, as "length of line," "length of single track" and "length of double track," when applied to street railways; and on account of such ambiguities the attempt has been made to fix upon some definite nomenclature by which careful returns can be secured. The conclusion is, that "length of line" means length of road-bed, or, in case of railways running entirely upon streets, the length of street occupied; that "length of single track" means the length of that portion of the road-bed or street laid with one track only; and that "length of double track" means the length of that portion of the road-bed or street laid with two tracks. In determining the total length of tracks, switches and sidings have been included, and thus double track has been reckoned as two tracks.

On December 31, 1889, 476 cities and towns in the United States possessed rapid-transit facilities, the total number of railways in independent operation being 807. Many railroads, however, (and the number is stated at 286, having a total length of 3,150.93 miles, and 13 having a total length of 135.75 miles), have as yet made no report; while in six the returns received were so imperfect that it was necessary to supplement them by approximations. This statement accounts for the bulletins not presenting statistics for a series of years for the whole number of railroads in the country, and 56 cities have been selected for which the reports are comparatively complete. Suburban lines tributary to large cities, but without their corporate limits, as well as those actually within the cities, are included in the statement; as, for instance, where cities situated close together have a common street railway system, it has not been thought best by the experts to attempt a separation in the tables. Therefore Pittsburg and Allegheny, in Pennsylvania, are treated as one city, as are also Newark and Elizabeth, in New Jersey. The street railway lines comprehended in Boston traverse also Lynn, Cambridge and other suburban places.

The aggregate mileage of the fifty-six cities selected for each year from 1880 to 1889, with the increase and percentage of increase, is shown in the following table:

YEAR.	TOTAL MILEAGE.	INCREASE.	
		MILES.	PER CENT.
1880	1,689.54	......	....
1881	1,765.95	76.41	4.52
1882	1,875.10	109.15	6.18
1883	1,941.49	66.39	3.54
1884	2,031.84	90.35	4.65
1885	2,149.66	117.82	5.80
1886	2,289.91	140.25	6.52
1887	2,597.16	307.25	13.42
1888	2,854.94	257.78	9.93
1889	3,150.93	295.99	10.37
Total	......	1,461.39	86.50

It is only fair to state that in order to make the foregoing statement, the statistics of some of the cities have been re-enforced by sources other than the census returns.

By the above table it will be seen that from 1,689.54, total mileage in the fifty-six cities selected in 1880, the growth has been to 3,150.93 miles in 1889. This is an increase of 1,461.39 miles, or 86.50 per cent. These figures show conclusively the rapidly increasing wants of cities.

The five leading cities of the country have a mileage assigned them as follows: Philadelphia, 283.47; Boston, 200.86; Chicago, 184.78; New York, 177.10; Brooklyn, 164.44. These are figures for 1889, and they show the total length of line; but the total length of all tracks, including sidings, for the same cities, is as follows: New York, 368.62; Chicago, 365.50; Boston, 329.47; Brooklyn, 324.63; Philadelphia, 324.21. From these figures we find that the position of Philadelphia in the last statement is reversed, and that New York steps from the fourth place in the five cities named to the first place; and this brings out a peculiarity of the Philadelphia roads and, to some extent, the roads of Boston, the tracks in these cities, to a large extent, occupying different streets in going to and from a terminus instead of being laid upon the same street.

The motive power used on the total mileage given is divided as follows:

MOTIVE-POWER.	MILES.	PER CENT.
Animal power	2,351.10	74.62
Electricity	260.36	8.26
Cable	255.87	8.12
Steam (elevated roads)	61.79	1.96
Steam (surface roads)	221.81	7.04
Total	3,150.93	100.00

The relative economy of cable, electric and animal motive power has been brought out by the census officers, but the superintendent remarks, in issuing the bulletins on this subject, that it is still too early to form a final judgment regarding the value of electric motive power for street railways; yet he feels that the statistics presented being, as they are, a record of actual experience, throw considerable light upon the matter of economy. The lack of uniform accounts of railways prevents the use of the data already collected for the formation of a final judgment; while again, the electric railways, being nearly all new, have not been in operation a sufficient length of time to afford final conclusions as to economy of service, and as Prof. Adams points out, most electric railways are the successors of roads operated by horses, the horses being still retained on a part of the lines and the expense incurred for horse-power being intermixed with that incurred for electric power. For these reasons a final judgment on the figures given must not be reached; yet the facts presented are indicative of what may be expected.

The bulletin relating to the relative economy of different motive powers embraces 50 lines of street railway, 10 of which are operated by cable, 10 by electricity, and 30 by animal power; and from the various tables presented, showing length, steepest grade, number of cars, car mileage, number of passengers carried, operating expenses, etc., a crystallized statement (which statement, it should be remembered, is not a complete and accurate one) is drawn, showing that the operating expense per car mile of cable railways is 14.12 cents; of electric railways, 13.21 cents; and of animal power, 18.16 cents;

while the operating expenses per passenger carried is, for cable railways, 3.22 cents; for electric railways, 3.82 cents; and for railways operated by animal power, 3.67 cents. It will surprise many to learn that in operation both cable and electric railways show a greater economy than railways operated by animal power; but in the full tables given in the bulletins it is noticeable that electric railways which have the least expense per car mile have the greatest expense per passenger carried. So the statement of the ratio between passengers carried and car mileage becomes essential, and from this it appears that electric railways show a less number of passengers per car mile that either of the other classes, the number of passengers carried per car mile being, for cable railways, 4.38; for electric railways, 3.46; and for railways operated by animal power, 4.95. Thus the electric railways carry a less number of passengers per car mile than either of those operated by cable or by animal power. The assumption is made in the census report that this variation is explained by the fact that electric roads, being new, occupy lines over which the passenger traffic has been but partly developed.

The expense per car mile and per passenger, the cost of road and equipment, and the volume of passenger traffic are essential for a full understanding of the financial side of the question. From the statistics reported it is seen that the total cost of road and equipment per mile of line (meaning thereby street length) is, for cable railways, $350,324.40; for electric railways, $46,697.59; and for railways operated by animal power, $71,387.38; and the number of passengers carried per mile per year is, for cable railways, 1,355,965; for electric railways, 222,648; and for railways operated by animal power, 596,563. From these figures it appears to be true that cable railways attain their greatest efficiency where an extremely heavy traffic is to be handled, and that electric railways and those operated by animal power are used where the traffic is not so heavy, or is more generally diffused.

The operating expense per car mile is: For cable railways, 14.12 cents; for electric railways, 13.21 cents; for railways operated by animal power, 18.16 cents; and the operating expense per passenger carried is, for the different powers as named, respectively, 3.22 cents, 3.82 cents, and 3.67 cents, but including interest charge per car mile at assumed rate of six per cent., the sum of operating expense and interest per car mile is: For cable railways, 20.91 cents; for electric railways, 17.56 cents; and for railways operated by animal power, 21.71 cents. These charges, both actual and estimated, show a somewhat greater expense for cable roads per car mile than for electric roads; but when the interest charge is considered on the basis of passengers carried, and added to the operating expense, the sum of operating expense and interest per passenger is as follows: For cable railways, 4.77 cents; for electric railways, 5.08 cents; for railways operated by animal power, 4.39 cents, showing a less cost for operating expense and interest charge per passenger for cable railways than for electric railways. In the first instance the greater charge for cable railways is on account of the much greater cost and equipment per mile, while the greater number of passengers carried by cable railways per mile reduces the ratio of expense on the passenger basis.

It is to be hoped that the complete statistics relating to rapid transit in cities will enable the public to determine, with reasonable accuracy, the relative economy of the different powers used. This is a question which is vital to the interest of city and suburban communities, and which leads to the ethical consideration of the problem of rapid transit. That power must eventually be used by which passengers can be transported from their homes to their places of business and return at the least possible expense and the greatest possible safety commensurate with high speed.

The necessity of living in sanitary localities, in moral and well-regulated communities, where children can have all the advantages of church and school, of light and air, becomes more and more evident as municipal governments undertake to solve the problems that are pressing upon them. If it be desirable to distribute the population of congested districts, through country districts, means must be provided for safe, rapid and cheap transit to the country districts; or inversely, if it be desirable to build up the suburban areas, the people must be supplied with cheap and convenient means of reaching the localities within which they earn their living.

The reduction of fares, through improved means of rapid transit, however desirable, is really a minor question. It is probably true, that by a slight reduction from a five-cent fare the head of a family engaged in mechanical labor, earning perhaps five or six hundred dollars per annum, might save enough to pay taxes, or to offset church and society assessments, or to furnish his family with boots and shoes, in any event extending his power *pro tanto* for the elevation of his family; but he does more than this when speed is taken into consideration. By the old methods of transit from suburbs to the heart of a city, a workingman going into the city of Boston was practically obliged, while working ten hours at his usual occupation, to spend an hour on the horse railway, when now, on one line, by the use of the electric car, he can go to and return from his place of work in half that time, thereby actually adding to his own time half an hour each day, practically reducing his working time from eleven hours to ten and a half hours without reduction of wages and without increased expense for transportation. The question of rapid transit, therefore, as seen by this simple illustration, becomes an ethical consideration; for if there is anything to be gained by adding to the time which men have at their disposal for their own purposes, for intercourse with their families, for social improvement, for everything for which leisure is supposed to be used, then the question of rapid transit is one of far greater importance than that of saving money either to the man who uses transportation or to the company that secures dividends upon its stock. I believe, therefore, that all efforts that are being made to secure convenient and cheap rapid transit in great cities are those which should bring to their support the help of all men who are seeking the improvement of the condition of the masses.

Business extension in cities is crowding the street area. This area is precisely the same in old cities like Boston, New York, Philadelphia, etc., for the present population and business operations that existed a century ago. The crowding of streets with the transportation essential for the movement of goods increases with great rapidity, and when the crowding is augmented, perhaps doubled, by the presence of the means of transporting passengers, the difficulties involved are almost appalling. With every increase of population the companies having in charge transportation facilities must, in order to accommodate the public, add more cars and more animals —if animals are the motive power—and so rapidly add to the already crowded condition of streets. This process is one which attacks the health and the safety of the people. The presence of so many horses constantly moving through the streets is a very serious matter. The vitiation of the air by the presence of so many animals is alone a sufficient reason for their removal, while the clogged condition of the streets impedes business, whether carried on with teams or on foot, and involves the safety of life and limb. It is a positive necessity, therefore, from this point of view alone, that the problems connected with rapid transit should be speedily solved, and this feature demands the efforts and the support of sanitarians. With the removal of tracks from the surface, and with tunnels built in such a manner as to be free from the dampness of the old form of tunnel, as has been done in London, and to secure light and air and be easy of access, all the unsanitary conditions of street railway traffic will be at once and forever removed; and if private capital cannot be interested to a sufficient extent to undertake such measures, then municipal governments must see to it that the health of the community is not endangered by surface traffic. When this question is allied to the ethical one, and when one considers the advantages to be gained, first, through securing rapid transit from the crowded portions of cities to the suburbs, and, second, by removing rapid transit traffic from the surface to underground viaducts, the importance of the whole problem becomes clearly apparent, and not only the importance of the problem but the necessity of its solution.

The statistics given by the census officers seem to indicate that as a matter of economy the very best equipment can be used without increasing the tax upon individual passengers. If underground roads can be used without at first increasing such tax, and still offer a reasonable compensation for capital invested, the gains to the people at large offer an inducement to capital, while the many considerations of health and morals offer men who desire to use their means for the benefit of their kind an opportunity that has not existed in the past. From my knowledge of some of the men who have been foremost in projecting lines of rapid transit, but who have been accused of doing it for entirely selfish motives, I learn that public benevolence has influenced them to a sufficient extent to induce them to take the great risks which are apparently involved. I believe that could the real, underlying patriotism of such men be known, and the confidence of the public in their willingness to do work for the public benefit gained, the solution of the rapid transit problem would be much easier.

Capital is securing less and less margin of profit through its investments, whether in manufacturing or in other enterprises. The capitalist is satisfied with a safe and sure return of from three to five per cent, and the spirit of altruism, which seems to be growing more and more rapidly among our millionaires, and which is leading them to the establishment of great institutions for public good, will lead them ultimately to such operations as those essential to secure the best results of rapid transit. Private capital, encouraged and protected by public sentiment and municipal enactments, may be capable of solving this problem. If it is not, then public sentiment, interested in the welfare of the people at large, not only from an economic point of view, but from sanitary and ethical considerations, will insist upon a public solution of the question. It is an important study, and the officers of the eleventh census are entitled to great credit for their efforts to bring out the partial results they have published, and later, to give to the country the full data relative to rapid transit in cities.

APPENDIX E.

ELECTRIC STREET RAILWAYS AS INVESTMENTS.*

BY LEMUEL WILLIAM SERRELL, M. E.

OF the forms of motive-power that have been tried to take the place of the horse may be mentioned the gas and compressed-air motors, the cable, the electric conduit, the storage battery and the trolley road. The gas and compressed-air systems are probably the oldest, and for the past twenty years they have been pushed by their advocates and put upon roads, on trial, all over this country and Europe; yet to-day there are no roads in operation by either motor — except experimentally — to show merits sufficient to cause their adoption.

There is no doubt that a suitable electric conduit will be invented some day, but we need no better instances of its failure in the past than the abandoned conduit in Fulton Street, New York, and the receipted bills of the West End Company, of Boston, for the sale of the old iron that had once been used for a similar purpose.

The history of the storage battery gives the same results. It has been favored because it is an ideal system. There is scarcely a large city in the country where storage-battery cars have not been run experimentally, and yet it has not been adopted because it has proved impracticable; while the trolley road, starting side by side with the storage battery, with all the maledictions that could be hurled upon it, has established itself as the only practical method of electric traction. The reported deadliness of the overhead wire has been proved a myth, and the objections to the system now are only æsthetic ones.

Let us look briefly at what has been done in the case of electric

* Abstract from the *Engineering Magazine* for May, 1892.

trolley railways. Scarcely five years have elapsed since it was shown that the trolley system could be made a practical success as a means of propelling cars, and yet to-day more than 450 roads are reported as being operated by electric power, having a total mileage of more than 3,600 miles and employing nearly 5,800 motor cars. Thus about three-eighths of the street railways in this country are now operated by the trolley system. The old tram rails are being replaced by better forms of construction, handsome cars measuring thirty feet in length replace the old style of horse cars, and a speed double that attainable with horses is used with perfect safety in equipping street roads with the trolley system. Many of our large cities are already so equipped, and it is estimated that $155,000,000 has already been expended. It has also been proposed that the experiment be tried to ascertain if electricity cannot be used practically to supersede steam on railways. Many of us doubtless will see this accomplished, although probably not until electricity can be generated directly from coal, without the use of the steam boiler, in which even a train of cars so propelled, it is estimated, will move at least five miles for the same cost that is now required to move a train of the same weight one mile by steam. Neighboring cities ten and fifteen miles apart have been connected together by such roads. A fifty-mile electric road is proposed between Worcester and Providence; another forty miles long, is being built between Tacoma and Seattle, and an electric road is projected between Chicago and St. Louis, to be built in a straight line, over which a speed of more than a hundred miles an hour is expected to be attained.

Thus we see developing from the old tram roads a system of roads operated with electric power which bids fair to be of as great, if not of greater importance to the world than the present steam railroads, which are a development from the same original tram road. The census of 1890 shows that the street railways of New York City carried during the year two thousand million passengers, or more than the entire population of the globe, and about four times as many passengers as were carried on the steam roads of the entire country, this number of passengers having been two hundred and fifty millions. The steam railway track mileage is about sixteen times as great as that of the street railway. The street car lines in the city of Boston, New York and Philadelphia carry more passengers per annum than all the steam roads of their respective States. Cable roads are usually built in crowded cities where it has become necessary to dispose of horses and the right for the overhead wire could not be obtained. No one would think for a moment, however, of building a cable road if a franchise could be obtained for a trolley road, as they cannot carry any more people, route for route, than the electric road, and cost nearly ten times as much to build.

The value of street railway securities as safe investments is only beginning to be generally appreciated. In the past they seldom have been offered to the public. The fact that a road once located upon a principal thoroughfare in a city is fixed upon the line of personal travel, which cannot be changed by the building of a parallel line upon another street, has not been brought forcibly enough before us to be generally appreciated. The fact is that a street railway located upon a principal thoroughfare, equipped with modern appliances for rapid transit, runs upon a highway through which people move, and they will ride on these cars rather than those of a parallel line on another street. The securities on such a property, when once on a paying basis, make a safer investment than steam railroad bonds, the value of which are always liable to be impaired by rate wars and by the building of parallel lines. Steam railroads run between terminals and not upon fixed lines of personal travel, and where terminal facilities can be secured in cities, parallel lines are always likely to be built.

When a city reaches a population of 25,000, its growth and business prosperity may safely be expected to increase steadily, and the securities on street railroads on a paying basis and located as above described, are sure to rise in value. The tendency of human nature is also to locate and build close to this line of travel, causing the thoroughfare to become more important as the city grows.

The introduction of electricity as a means of rapid transit has done, and is doing much to bring this class of securities before the investing public. . . .

The remarkable cheapness with which electric roads can be operated as compared with horse roads and the cheapness of the first cost as compared with cable roads has led to the building of more than 3,600 miles of such roads within the past five years. The earnings for the capital invested are larger for electric street railways than for steam, horse or cable roads, and the securities on such properties are now beginning to attract the attention of the general public. The cost of building and equipping street roads varies considerably with different local conditions. A comparison of the average cost of building and equipping cable, electric and horse railways per mile of track, taken from the recent census reports, is given below :

```
Cable roads.....................................................$350,000
Electric roads.........................................  .............  30,000
Horse roads....................................................  41,000
```

The above represents a fair average for paved streets in cities proper. For suburban travel the cost per mile of track and electric equipment need not exceed $20,000.

The earnings of the various properties may best be expressed by the ratio of their operating expenses to their gross receipts. It is hard to get figures giving this ratio for cable roads. The only ones published that I have been able to find are for Denver, Col., where this ratio is reported as 77 per cent, while for the electric lines owned by the same company it is reported at 55 per cent. Cable roads cost almost as much as elevated railroads, yet in some places they both are bonded for more than $1,000,000 per mile and have dividend stocks. The same ratio for horse railroads we find, by taking the average of over fifty roads from the reports of railroad commissioners, is 80 per cent.

A large number of reports have been published on electric roads, placing this ratio at about 50 per cent. From my own investigation I have found that these figures were true for certain months, but they do not represent the average for the whole year, and that in the section of the country where snow must be removed from the tracks the year's average is from 60 per cent to 65 per cent. The increase in the net earnings in cities like Boston per revenue car mile run for electric road over horse roads averages 50 per cent. This enormous increase of earning capacity has given a well merited "boom" to the electric railway, but at the same time has opened an opportunity for objectionable speculation, the same as surrounded the steam railroads, which caused the wrecking of so many fortunes and gave the opportunity for great railroad steals, followed by reorganizations and "freeze-outs."

* * * * * *

Electric railway securities are comparatively a new thing, and being so little known are looked upon as not being desirable. The skepticism that has surrounded the mysterious action of electricity has not furnished the same fertile basis upon which to float securities as the development of commerce by building steam railroads. Therefore the people who have attempted to make more than their share of profit by aping the plan upon which steam roads were built probably will have a heavy burden to carry for some time. But there are elements of security in a well-selected street railway bond which make it as good an investment as any water works or municipal bond and safer than steam railways.

Bonds on electric railways may be divided into three classes:

1. Bonds on new roads built from franchises, whose earnings are problematical.

2. Bonds on reorganized horse roads, whose net earnings with horses have not been sufficient to pay the interest on their bonded debt when equipped with electricity.

3. Bonds on reorganized horse roads, whose net earnings with electricity are sufficient to pay the interest on their bonded debt when equipped with electricity.

Projectors of a road of the first class need not expect to sell their securities, except at a great sacrifice, until the road shall have been in operation long enough to demonstrate its earning capacity, and even then the securities should not be purchased without careful personal examination and endorsement by responsible parties.

Bonds on roads of the second class may be considered as very good and safe investments, where the interest charges on the bonds do not exceed by more than one-fourth the net earnings of the roads when operated with horses. Capitalists will run very little risk in undertaking the financiering of roads of this class when the above restriction is observed and the cities are of good size and well known. The stock on such roads probably will be dividend stocks from the very start.

Bonds on roads of the third class are of the best class and are not excelled by any other form of investment. Such bonds should sell at a premium and find a very ready market, and the stock should be worth par.

When the elements of security that surround street railway bonds become better appreciated they will be regarded among the best and most readily negotiable securities before the public. In cities from 25,000 to 40,000 people a comparatively local market must be sought, but well-selected bonds on roads in cities having a larger population should find a ready sale with the general public.

INDEX.

ARMATURES, causes that make it revolve, 10.
        ring, 15.
        troubles with, what to do, 128.
        removing dust from, 133.
Ampere-turns, 17.
Ammeter, 14.
Alkaline zincate, 106.
Accumulator, forms of, 104.
        electro-motive force of, 106.
        internal resistance of, 106.
        alkaline zincate, 106.
        lead, 104-106.

BATTERIES, storage, 104-106.
Boilers, as arranged by the West End Company, Boston, Mass., 135.
Boiler house, 135.
Brushes, sparking at, 128.
        neutral points of, 129.
Burton Car Heater, 75-78.

CARS, heating of, 75-78.
        Ford and Washburn Storage, 106-109.
        seating capacity, 123.
        wiring of, 98-103.
Clamping cars, 33.
Cables for car wiring, 98.
Carbon brushes, 129.
Carbon dust, 129.
Cut-outs, automatic, 9.
Cut-out switches, 99.
Controlling switches, 99.
Commutator, 133.
        trouble at, 128.

INDEX.

Current, path of, in car circuit, 103.
    methods of distributing on the line, 7-10.
    indicator, 134.
Crosses, for trolley wire, 33.
Collisions, how avoided, 127.
Contact, imperfect, 125.

DROP of potential, 134.
Dynamos, 15.
    railway, 15-25.
    shunt-wound, 15.
    series-wound, 15.
    compound-wound, 17.
    Thomson-Houston Railway, 19.
    The Mather Railway, 21.
    Westinghouse multipolar, 21-23.
    Edison Railway, 25.
    Short railway, 17-19.
    care of, 132-133.

ELECTRIC railway, 145.
    history of, 145-148.
    as investments, 176.
Electric plant, 135.
    West End, Boston, Mass., 135-136.
Engine, 136.
    Reynolds & Corliss, 136.
Electro-motive force, 15.
    amount of, for railway work, 15.

FREIGHT locomotive, 84.
    Thomson-Houston, 84-89.
    The Edison, 89-91.
Fenders, 149.
    experiments with, 149-150.
Feeders, 9.
    use of, 34-35.
Field Circuit, 15-17.
    magnets, 15.
Force, electro-motive, 15.
Frogs, 33.
    for line construction, 33.

GUARD wires, 36.

Gear, 41.
    use of, 41.

HEATING, 71.
    of rheostat, 71.
    of cars, 75-78.
    of fields, 130.
    of ammeters, 134.

INDICATORS, 14.
    of currents, 134.
Insulators, of tools, 134.
    of trolley wire, 29-32.
    difficulties of good, 98.
    of base of dynamos, 132.
    oil as an insulator, 133.
Imperfect contact, 125.
    causes of, 125.

LOCOMOTIVE, 84.
    for heavy traction, 84.
    new 100 h. p. Thomson-Houston freight, 84-89.
    The Edison mining, 89-91.
Lightning arrester, station, 11.
    on car, 125.
    grounded, 125.
Line, 7.
    construction of, 26-40.
    plan of single curve overhead, 38.
    plan of double curve overhead, 39.

MAGNETIC field, 15.
Magnetic effect, 15.
Motor, electric, 41.
    construction of, 41.
    Wightman single railway, 41-47.
    Thomson-Houston single reduction gear railway, 47-50.
    Thomson-Houston waterproof, 51-53.
    Westinghouse four-pole single, 53-61.
    Edison slow speed single rod, 61-63.
    Short gearless, 63-69.
    arrangement of field spools, 126-127.
    street car, 41-69.
    difficulties with, 128-129.

POWER station, 11.
    plan for, small, 12.
Poles, 26.
    construction of, 26.
    size, 27.
    arrangements of, 27.
    wooden, 27-30.

RAILROADING, electric, 138.
    safety of, 138-143.
    high-speed, 144.
    history of, 145-148.
Rheostat, 70.
    resistance of, 70.
    heating, 70.
    field, 71.
    car, 71.
    Thomson-Houston Railway, 72-74.
    Short Electric Railway, 74.
Resistance of rheostat, 70.
Roads, 110.
    illustrative, 110-123.
    rapid transit, 110-116.
    experiments of storage battery on, 106-109.
Rails for electric lines, 36-37.
    objections to using them for ground circuit, 40.

STEAM engine, 136.
    Reynolds & Corliss, 136.
Storage battery, 104-106.
    car, 107.
    Ford & Washburn's, 106-109.
Street railway, 123.
    dynamos for, 15-25.
    West End, Boston, 123-136.
    Lynn & Boston, 123.
Street car motors, methods of controlling, 151.

TRANSIT, rapid, 110, 116, 144.
Trolley, described, 9-10.
    the wire, 9.
    apparatus, 79-83.
    poles, 79-83.

Trolley, The Rae, 79.
    The Boston, 81-82.
    The Baker, 81-82.
    Common Sense, 81-82.
    The Short Sliding, 82.
    old form of, 81.
    The Wightman, 83.
Tracks, 36-37.
Traction, 84.
    locomotives by heavy, 84.
    by storage batteries, 106-109.
Trucks, 92-97.
    The Rae, 94.
    The Radial Car, 94.
    The Bogie, 94.
    The Brill Car Co., 97.

WINDING, 15.
    for series dynamos, 15.
    for compound dynamos, 17.
    for shunt dynamos, 15.

www.ingramcontent.com/pod-product-compliance
Lightning Source LLC
Chambersburg PA
CBHW032145160426
43197CB00008B/781